D1082911

ON NUMBERS AND GAMES

SECOND EDITION

ON NUMBERS AND GAMES

SECOND EDITION

J. H. CONWAY

John von Neumann Professor of Mathematics,
Princeton University

A K Peters, Ltd.
Natick, Massachusetts

Editorial, Sales, and Customer Service Office

A K Peters, Ltd.
63 South Avenue
Natick, MA 01760

Library of Congress Cataloging-in-Publication Data

Conway, John Horton.
On numbers and games / John H. Conway.–2nd ed.
p.cm.
Includes bibliographical references and index.
ISBN 1-56881-127-6 (alk. paper)
1. Number theory. 2. Game theory. I. Title.

QA241 .C69 2000
519.3–dc21 00-046927

Printed in the United States of America
05 04 03 02 01 10 9 8 7 6 5 4 3 2 1

Prologue

Just over a quarter of a century ago, for seven consecutive days I sat down and typed from 8:30 am until midnight, with just an hour for lunch, and ever since have described this book as "having been written in a week."

Not entirely honest, because there were loose ends still to be tied up, and Chapter 16 was written just before the book appeared, while Chapter 13 was largely copied from a paper, "Hackenbush, Welter and Prune", that had been written a year earlier. But also not entirely dishonest.

Why the rush? Because ONAG, as the book is familiarly known, was getting in the way of writing *Winning Ways* (WW). Now that both books are happily being republished by A K Peters, Onagers (a word that also means "Wild Asses"!) can be told just how it came about before they surrender themselves to pure pleasure (as "Onag" means in Hebrew!).

A few years previously, Elwyn Berlekamp, Richard Guy and I had agreed to write a book on mathematical games, by which at that time we meant the Nim-like theory developed independently by Roland Sprague and Peter Michael Grundy for sums of impartial games—those for which the two players have exactly the same legal moves.

I had long intended to see what would become of the theory when this restriction was dropped, but only got around to doing so when the then British Go Champion became a member of the Cambridge University Pure Mathematics Department. Astonishingly, it was the resulting attempt to understand "Go" that led to the discovery of the Surreal Numbers! This happened because the typical "Go" endgame was visibly a sum of games in the sense of this book, making it clear that this notion was worthy of deep study in its own right. The Surreal Numbers then emerged as the simplest domain to which it applies!

However, their theory rapidly burgeoned in ways that made it inappropriate for the book that later became *Winning Ways*. A busy term was approaching, and it seemed that this "transfinite" material just had to be got out of the way before that term started if *Winning Ways* was ever to be published. So I sat down for that week and wrote this book, and then confessed the fact to my co-authors.

The most surprising immediate result was a threat of legal action from Elwyn Berlekamp! But somehow we must have patched this up, because both ONAG and WW appeared in the next few years, and we remain good friends.

In fact, the Surreal Numbers "surfaced" before ONAG appeared, partly through my 1970 lectures at Cambridge and Cal. Tech., but mostly through the wide circulation of Donald Knuth's little book, *Surreal Numbers*. I am very grateful to Knuth for inventing this name—the original version of ONAG said "Because of the generality of this Class, we shall simply describe its members as numbers, without adding any restricting adjective." "Surreal Numbers" is much better!

I am very happy and grateful that A.K. Peters have agreed to publish millennial editions of both this book and *Winning Ways*.

Ariel Jaffee and Kathryn Maier were responsible for handling the changes to this edition. This is also the place to acknowledge Richard Guy's considerable contributions to the original edition. In particular, he designed and drew a number of the original figures and computed, or recomputed several of the tables.

I have called this a Prologue rather than a Preface because it is usually understood that the Preface to a later edition of a book should contain a description of the changes in the book and its subject since its first edition. Some of these functions are addressed in the Epilogue.

John H. Conway

Preface

This book was written to bring to light a relation between two of its author's favourite subjects—the theories of transfinite numbers and mathematical games. A few connections between these have been known for some time, but it appears to be a new observation that we obtain a theory at once simpler and more extensive than Dedekind's theory of the real numbers just by *defining* numbers as the strengths of positions in certain games. When we do this the usual properties of order and the arithmetic operations follow almost immediately from definitions that are naturally suggested, so that it was quite an amusing exercise to write the zeroth part of the book as if these definitions had arisen instead from an attempt to generalise Dedekind's construction!

However, we suspect that there will be many readers who are more interested in playing games than philosophising about numbers. For these readers we offer the following words of advice. Start reading Chapter 7, on playing several games at once, and find an interested friend with whom to play a few games of the domino game described there. In this it's easy to

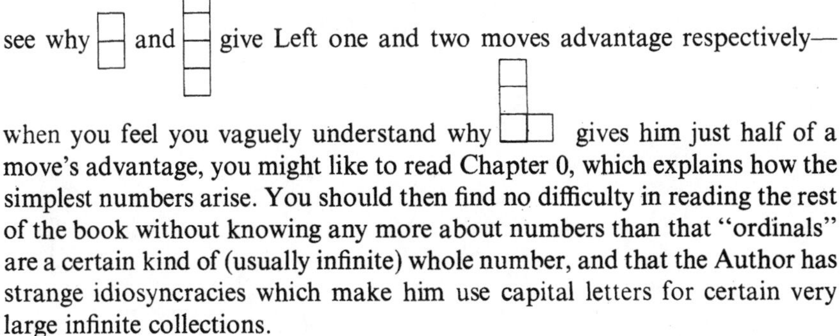

see why and give Left one and two moves advantage respectively—when you feel you vaguely understand why gives him just half of a move's advantage, you might like to read Chapter 0, which explains how the simplest numbers arise. You should then find no difficulty in reading the rest of the book without knowing any more about numbers than that "ordinals" are a certain kind of (usually infinite) whole number, and that the Author has strange idiosyncracies which make him use capital letters for certain very large infinite collections.

Many friends have helped me to write this book, often without being aware of the fact. I owe an especial debt to Elwyn Berlekamp and Richard Guy, with whom I am currently preparing a more extended book on mathematical games which should overlap this one in several places. The book would never have appeared without the repeated gentle proddings that came from Anthony Watkinson of Academic Press; it would have contained many errors were it not for the careful reading of Paul Cohn as editor, and the quality of the printing and layout could never have been so high without the detailed attentions of Ron Hitchings and the staff of the printers at Page Bros of Norwich. Others whose comments have affected more than one page are Mike Christie, Aviezri Fraenkel, Mike Guy, Peter Johnstone, Donald Knuth and Simon Norton. The varied nature of the advice they gave is neatly encapsulated in the following lines from Bunyan's *Apology for his Book (Pilgrim's Progress)*:

> *Some said 'John, print it'; others said, 'Not so.'*
> *Some said 'It might do good'; others said, 'No.'*

October 1975 J.H.C.

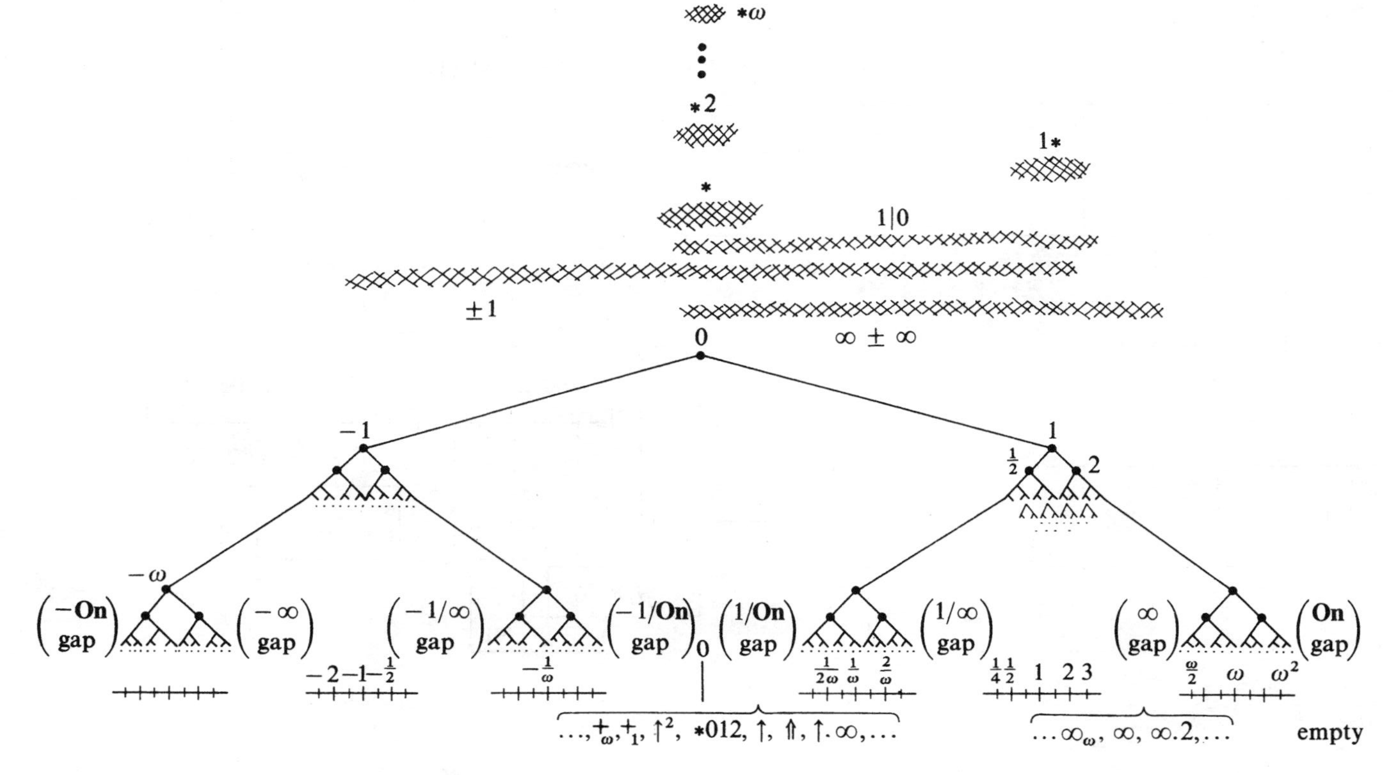

Frontispiece. The tree of numbers, and the positions of some games.

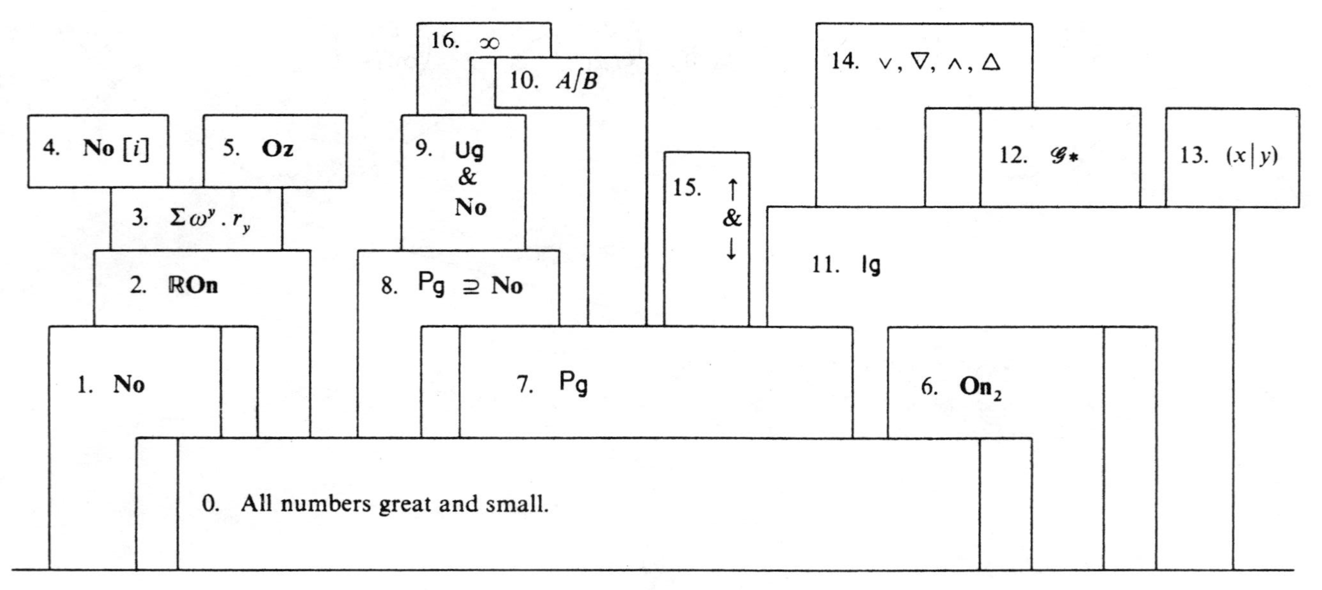

Dependence of Chapters. (Not on oath!)

Contents

ZEROTH PART

ON NUMBERS . . .

A Hair, they say, divides the False and True;
Yes; and a single Alif were the clue,
Could you but find it—to the Treasure-house,
And peradventure to The Master too!

Edward Fitzgerald's
"Rubaiyat of Omar Khayyam"

CHAPTER 0

All Numbers Great and Small

Whatever is not forbidden, is permitted.
J. C. F. von Schiller, Wallensteins Lager

This book is in two = {zero, one | } parts. In this zeroth part, our topic is the notion of *number*. As examples we have the finite numbers $0, 1, 2, \ldots$, $-1, \frac{1}{2}, \sqrt{2}, \pi, \ldots$; infinite numbers such as ω (the first infinite ordinal); and also infinitesimal numbers such as $1/\omega$. If we were to adopt the axiom of choice, then the infinite cardinal numbers like $\aleph_0$ could be identified with the least corresponding ordinal numbers, so that we can regard these too as part of our system (although the arithmetic is different).

In the system of "Surreal Numbers" we shall describe, every number has its own unique name and properties and many remarkable numbers, such as

$$\sqrt[3]{(\omega + 1)} - \frac{\pi}{\omega}$$

appear. But the "number" $i = \sqrt{-1}$ will not arise in the same way (though we add it in Chapter 4), since there is no property enjoyed by i which is not shared by $-i$. In fact we reply to questions about "the square root of -1" by simply asking exactly which square root of -1 is meant?

Let us see how those who were good at constructing numbers have approached this problem in the past.

Dedekind (and before him the author—thought to be Eudoxus—of the fifth book of Euclid) constructed the real numbers from the rationals. His method was to divide the rationals into two sets L and R in such a way that no number of L was greater than any number of R, and use this "section" to define a new number $\{L \mid R\}$ in the case that neither L nor R had an extremal point.

His method produces a logically sound collection of real numbers (if we ignore some objections on the grounds of ineffectivity, etc.), but has been criticised on several counts. Perhaps the most important is that the rationals are supposed to have been already constructed in some other way, and yet are "reconstructed" as certain real numbers. The distinction between the "old" and "new" rationals seems artificial but essential.

Cantor constructed the infinite ordinal numbers. Supposing the integers $1, 2, 3, \ldots$ given, he observed that their *order-type* ω was a new (and infinite) number greater than all of them. Then the order-type of $\{1, 2, 3, \ldots, \omega\}$ is a still greater number $\omega + 1$, and so on, and on, and on. The similar objections to Cantor's procedure have already been met by von Neumann, who observes that it is unnecessary to suppose $1, 2, 3, \ldots$ given, and that it is natural to start at 0 rather than 1. He also takes each ordinal as the *set* (rather than the order-type) of all previous ones. Thus for von Neumann, 0 is the empty set, 1 the set $\{0\}$, 2 the set $\{0, 1\}, \ldots, \omega$ the set $\{0, 1, 2, \ldots\}$, and so on.

In this chapter we shall show that these two methods are part of a simpler and more general one by which we can construct the very large Class **No** of "Surreal Numbers," which includes both the real numbers and the ordinal numbers, as well as others like those mentioned above. Inside this book we shall usually omit the adjective "surreal," coined by Donald Knuth, and simply call these things "numbers." It turns out that **No** is a Field (i.e., a field whose domain is a proper Class)—in general we shall capitalise the initial letter of any "big" concept, on the grounds that proper Classes, like proper names, deserve capital letters. So, for instance, the word *Group* will mean any group whose domain is a proper class.

CONSTRUCTION

If L, R are any two sets of numbers, and no member of L is $\geqslant$ any member of R, then there is a number $\{L \mid R\}$. All numbers are constructed in this way.

CONVENTION

If $x = \{L \mid R\}$ we write x^L for the typical member of L, and x^R for the typical member of R. For x itself we then write $\{x^L \mid x^R\}$.

$x = \{a, b, c, \ldots \mid d, e, f, \ldots\}$ means that $x = \{L \mid R\}$, where $a, b, c, \ldots$ are the typical members of L, and $d, e, f, \ldots$ the typical members of R.

DEFINITIONS

Definition of $x \geqslant y$, $x \leqslant y$.
We say $x \geqslant y$ iff (no $x^R \leqslant y$ and $x \leqslant$ no y^L), and $x \leqslant y$ iff $y \geqslant x$.
We write $x \nleqslant y$ to mean that $x \leqslant y$ does not hold.

Definition of $x = y$, $x > y$, $x < y$.
$x = y$ iff ($x \geqslant y$ and $y \geqslant x$). $x > y$ iff ($x \geqslant y$ and $y \ngeqslant x$).
$x < y$ iff $y > x$.

Definition of $x + y$.
$$x + y = \{x^L + y, x + y^L \mid x^R + y, x + y^R\}.$$

Definition of $-x$.
$$-x = \{-x^R \mid -x^L\}.$$

Definition of xy.
$$xy = \{x^Ly + xy^L - x^Ly^L, x^Ry + xy^R - x^Ry^R \mid x^Ly + xy^R - x^Ly^R, x^Ry + xy^L - x^Ry^L\}.$$

It is remarkable that these few lines already define a real-closed Field with a very rich structure.

We now comment on the definitions. A most important comment whose logical effects will be discussed later is that *the notion of equality is a defined relation.* Thus apparently different definitions will produce the same number, and we must distinguish between the *form* $\{L \mid R\}$ of a number and the number itself.

All the definitions are inductive, so that to decide, for instance, whether $x \geqslant y$ we must consider a number of similar questions about the pairs x^R, y and x, y^L. but these problems are all simpler than the given one. It is perhaps not quite so obvious that the inductions require no basis, since ultimately we are reduced to problems about members of the empty set.

In general when we wish to establish a proposition $P(x)$ for all numbers x, we will prove it inductively by deducing $P(x)$ from the truth of all the propositions $P(x^L)$ and $P(x^R)$. We regard the phrase "all numbers are constructed in this way" as justifying the legitimacy of this procedure. When proving propositions $P(x, y)$ involving two variables we may use *double induction*, deducing $P(x, y)$ from the truth of all propositions of the form $P(x^L, y)$, $P(x^R, y)$, $P(x, y^L)$, $P(x, y^R)$ (and, if necessary, $P(x^L, y^L)$, $P(x^L, y^R)$, $P(x^R, y^L)$, $P(x^R, y^R)$). Such multiple inductions can be justified in the usual way in terms of repeated single inductions.

We shall allow ourselves to use certain expressions $\{L \mid R\}$ that are not numbers, since they do not satisfy the condition that no member of L shall be $\geq$ any member of R. In general we may write down any expression $\{L \mid R\}$ and even discuss inequalities between such expressions before establishing that they are numbers, but if we wish such an expression to represent a number we must establish the condition on L and R. In the more general theory developed in the next part of the book, we show that when the condition on L and R is omitted we obtain the more general notion of a *game*.

Our next comments concern the motives for these particular definitions. Now it is our intention that each new number x shall lie between the numbers x^L (to the left) and x^R (to the right), and that $\geqslant, +, -, .$, etc., shall have their usual properties. So that if (say) $y \geqslant$ some x^R we would not have $x \geqslant y$, for then $x \geqslant x^R$. Similarly, we could not allow $x \geqslant y$ if $x \leqslant$ some y^L. So we define $x \geqslant y$ in all other cases. (This conforms with our motto, and helps to ensure that numbers are totally ordered.)

The spirit of the definitions is to ask what we know already (i.e. by the answers to *simpler* questions) about the object being defined, and to make the answers part of our definition. Thus if addition is to have nice properties and if x is between x^L and x^R, and y between y^L and y^R, then we know "already" that $x + y$ must lie between both $x^L + y$ and $x + y^L$ (on the left) and $x^R + y$ and $x + y^R$ (on the right), which yields the definition of $x + y$. Similarly $-x$ will lie between $-x^R$ (on the left) and $-x^L$ (on the right), which suffice to define $-x$.

It is not nearly so easy to find exactly what we "already" know about xy. It might seem, for instance, that we know that xy lies between x^Ly and xy^L (on the left) and x^Ry and xy^R (on the right), which would yield the definition

$$xy = \{x^Ly, xy^L \mid x^Ry, xy^R\}.$$

But this fails in two ways. Firstly, what we "knew" here is sometimes false (consider negative numbers), and secondly, even when it is true it need not be the strongest information we "already" know. In fact, of course, this defines the same function as $x + y$.

It takes a great deal of thought to find the correct definition, which comes from the observation that (for instance) from $x - x^L > 0$ and $y - y^L > 0$ we can deduce $(x - x^L)(y - y^L) > 0$, so that we must have $xy > x^Ly + xy^L - x^Ly^L$. Since all the products here are simpler ones, and since we regard addition and subtraction as already defined, we can regard this inequality as already known when we come to define xy, and the other inequalities in the definition are similar. [Note that for positive numbers x and y the inequality $xy > x^Ly + xy^L - x^Ly^L$ is stronger than both inequalities $xy > x^Ly$, $xy > xy^L$.]

We can summarise our comments by saying that the definitions of the various operations and relations are just the simplest possible definitions that are consistent with their intended properties. In the next chapter, we shall verify that these intended properties really hold of all numbers, but in the rest of this chapter we shall simply explore the system in a more informal way. To simplify the notation, when L is the set $\{a, b, c, \dots\}$ and R the set $\{\dots, x, y, z\}$, we shall simply write $\{a, b, c, \dots \mid \dots, x, y, z\}$ for $\{L \mid R\}$.

EXAMPLES OF NUMBERS, AND SOME OF THEIR PROPERTIES

The number 0

According to the construction, every number has the form $\{L \mid R\}$, where

L and R are two sets of earlier constructed numbers. So how can the system possibly get "off the ground", since initially there will be no earlier constructed numbers?

The answer, of course, is that even before we have any numbers, we have a certain *set* of numbers, namely *the empty set* $\varnothing$! So the earliest constructed number can only be $\{L \mid R\}$ with both $L = R = \varnothing$, or in the simplified notation, the number $\{\,|\,\}$. This number we call 0.

Is 0 a number? Yes, since we cannot have any inequality of the form $0^L \geqslant 0^R$, for there is neither a 0^L nor a 0^R!

Is $0 \geqslant 0$? Yes, for we can have no inequality of the form $0^R \leqslant 0$ or $0 \leqslant 0^L$. So by the definition, and happily, we have $0 = 0$. We also see from the definitions that $-0 = 0 + 0 = 0$, since there is no number of any of the forms $-0^R, -0^L, 0^L + 0, 0 + 0^L, 0^R + 0, 0 + 0^R$. A slightly more complicated observation of the same type is that $x0 = 0$, since in every one of the terms defining xy there is a mention of y^L or y^R, so that when $y = 0$ no term is needed and the expression for xy reduces to $\{\,|\,\} = 0$. So the number 0 has at least some of the properties we know and love.

The numbers 1 and -1

We can now use the sets $\{\,\}$ and $\{0\}$ for L and R, obtaining hopefully the numbers $\{\,|\,\}$, $\{0\,|\,\}$, $\{\,|\,0\}$, $\{0\,|\,0\}$. But since we have already proved that $0 \geqslant 0$, $\{0\,|\,0\}$ is *not* a number, and we have only two new cases, which we call $1 = \{0\,|\,\}$ and $-1 = \{\,|\,0\}$. Note that -1 is indeed a case of the definition $-x$.

Is $0 \geqslant 1$? This will be true unless there is 0^R with $0^R \leqslant 1$ (there isn't) or 1^L with $0 \leqslant 1^L$ (there is, namely $1^L = 0$). So we do *not* have $0 \geqslant 1$.

Is $1 \geqslant 0$? This is true unless there is 1^R with "..." or 0^L with "..." (whatever "..." is, there plainly can't be). So we have $1 \geqslant 0$, and so $1 > 0$.

By symmetry, we have $-1 < 0$, and so if inequalities "behave", then we should have $-1 < 1$. We check this:

Is $-1 \geqslant 1$? This will happen unless there is $(-1)^R \leqslant 1$ or ... (there is, namely $(-1)^R = 0$). So we do not have $-1 \geqslant 1$.

Is $1 \geqslant -1$? This will happen unless there is 1^R with ... or $(-1)^L$ with ... (there isn't). So $1 \geqslant -1$, so $1 > -1$, as we hoped.

We can generalise a part of this last argument. If there is no x^R and no y^L, then $x \geqslant y$ holds vacuously.

We forgot to check that $1 \geqslant 1$. Why not do this yourself?

The numbers 2, $\frac{1}{2}$, and their negatives

We now have three numbers $-1 < 0 < 1$, and so a whole battery of particular sets

$$\{\,\}, \{-1\}, \{0\}, \{1\}, \{-1, 0\}, \{-1, 1\}, \{0, 1\}, \{-1, 0, 1\}$$

to use for L and R. But the condition that no member of L should be $\geqslant$ any member of R restricts us to the possibilities

$$\{\,|\,R\}, \{L\,|\,\}, \{-1\,|\,0\}, \{-1\,|\,0, 1\}, \{-1\,|\,1\}, \{0\,|\,1\}, \{-1, 0\,|\,1\}.$$

If our hopes are fulfilled, we should have $\{1\,|\,\} > 1$ and $0 < \{0\,|\,1\} < 1$. So we anticipate their probable values, and define $\{1\,|\,\} = 2$, $\{0\,|\,1\} = \frac{1}{2}$. We then have, of course, $\{\,|\,-1\} = -2$, and $\{-1\,|\,0\} = -\frac{1}{2}$, from the definition of negation.

Before we justify these names, let us ask about some of the other possibilities. For example, what about the number $x = \{0, 1\,|\,\}$? This x is presumably restricted by the conditions $0 < x$, $1 < x$. But since $0 < 1$, if inequalities behave (and we shall suppose from now on that they do), the condition $1 < x$ already implies $0 < x$, so in some sense the entry 0 isn't telling us anything. We can therefore hope that $x = \{0, 1\,|\,\} = \{1\,|\,\} = 2$. We test this, supposing $2 > 1 > 0$.

Is $x \geqslant 2$? This is so unless there is $x^R \leqslant 2$ (no) or $x \leqslant$ some 2^L (no, because the only 2^L is 1, and we believe $x > 1$). So we think that $x \geqslant 2$.

Is $2 \geqslant x$? Yes, unless some $2^R \ldots$ (no) or $2 \leqslant$ some x^L (no, since the only x^L are 1 and 0). So indeed $x = 2$, if all our expectations are fulfilled.

In a similar way, we should expect all the equalities in the table:

$$\begin{aligned}
-2 &= \{\,|\,-1\} = \{\,|\,-1, 0\} = \{\,|\,-1, 1\} = \{\,|\,-1, 0, 1\}\\
-1 &= \{\,|\,0\} = \{\,|\,0, 1\}\\
-\tfrac{1}{2} &= \{-1\,|\,0\} = \{-1\,|\,0, 1\}\\
0 &= \{\,|\,\} = \{-1\,|\,\} = \{\,|\,1\} = \{-1\,|\,1\}\\
\tfrac{1}{2} &= \{0\,|\,1\} = \{-1, 0\,|\,1\}\\
1 &= \{0\,|\,\} = \{-1, 0\,|\,\}\\
2 &= \{1\,|\,\} = \{0, 1\,|\,\} = \{-1, 1\,|\,\} = \{-1, 0, 1\,|\,\}.
\end{aligned}$$

Clearly we need some way of automating our expectations. Let us ask when the number $X = \{y, x^L\,|\,x^R\}$ obtained by adding a new entry y to the left of x is still equal to x.

Is $X \geqslant x$? Yes, unless some $X^R \leqslant x$ (no, since every X^R is an x^R) or $X \leqslant$ some x^L (no, since every x^L is an X^L).

Is $x \geqslant X$? Yes, unless some $x^R \leqslant X$ (no, since every x^R is an X^R) or $x \leqslant$ some X^L (and so $x \leqslant y$, since every other X^L is an x^L). We conclude that provided $y \ngeqslant x$, we can add y to the left of x in this way without affecting

x. This justifies all the equalities in the table. (We allow also, of course, y to be inserted on the right if $y \nleqslant x$.)

[In the case $\{-1 \mid 1\}$ we need to use the process twice. Thus since $-1 \ngeqslant 0 = \{\,\mid\,\}$, we have $0 = \{-1 \mid \,\}$. Then since $1 \nleqslant 0 = \{-1 \mid \,\}$, we have $0 = \{-1 \mid 1\}$.]

It is not hard to check the inequalities

$$-2 < -1 < -\tfrac{1}{2} < 0 < \tfrac{1}{2} < 1 < 2,$$

which shows that at least these numbers have the right order properties. What else do we require to justify their names?

According to the definition

$$1 + 1 = \{0 + 1, 1 + 0 \mid \,\},$$

since 0 is the only 1^L, and there is no 1^R. So provided $0 + 1$ and $1 + 0$ behave as expected, we have $1 + 1 = 2$, as we might hope. But provided $x^L + 0 = x^L$ and $x^R + 0 = x^R$, we have

$$x + 0 = \{x^L + 0 \mid x^R + 0\} = \{x^L \mid x^R\} = x,$$

and similarly $0 + x = x$. Since we already know $0 + 0 = 0$, this shows that $1 + 0 = 0 + 1 = 1$, as we wanted for the proof of $1 + 1 = 2$, but in fact it gives us an inductive proof that $x + 0 = 0 + x = x$ for all x.

It is much harder to show that $\frac{1}{2} + \frac{1}{2} = 1$, justifying the name of $\frac{1}{2}$. From the definition (supposing that $x + y = y + x$ for all x, y, which is quite easy to prove inductively) we see that

$$\tfrac{1}{2} + \tfrac{1}{2} = \{\tfrac{1}{2} \mid 1\tfrac{1}{2}\},$$

where we are using $1\frac{1}{2}$ as a temporary name for $1 + \frac{1}{2}$.

Is $\frac{1}{2} + \frac{1}{2} \geqslant 1$? Yes, unless $1\frac{1}{2} \leqslant 1$ or $\frac{1}{2} + \frac{1}{2} \leqslant 0$. Oh my, we have to do these first. Let's get on with it.

Is $1 \geqslant 1\frac{1}{2}$? Yes, unless (empty) or $1 \leqslant$ some $1\frac{1}{2}^L$. But one of the $(1 + \frac{1}{2})^L$ is $1 + 0 = 1$, so $1 \ngeqslant 1\frac{1}{2}$.

Is $0 \geqslant \frac{1}{2} + \frac{1}{2}$? Yes, unless (empty) or $0 \leqslant$ some $(\frac{1}{2} + \frac{1}{2})^L$. But since $0 \leqslant \frac{1}{2} + 0$, we have $0 \ngeqslant \frac{1}{2} + \frac{1}{2}$. So (at last) $\frac{1}{2} + \frac{1}{2} \geqslant 1$.

Now is the time to leave the question

$$\text{“is } 1 \geqslant \tfrac{1}{2} + \tfrac{1}{2}\text{?”}$$

to the reader. He should conclude that indeed $\frac{1}{2} + \frac{1}{2} = 1$.

In most of our examples x^L and x^R have been fairly close to each other, so that there was an obvious candidate for $\{x^L \mid x^R\}$. When they are far apart, there will be many simple numbers in between—which one of these will $\{x^L \mid x^R\}$ be? We consider $x = \{-1 \mid 2\}$.

Is $x \geqslant 0$? Yes, unless $2 \leqslant 0$ (false) or $x \leqslant$ some 0^L (false). So in this case we have $x \geqslant 0$.

Is $0 \geqslant x$? Yes, unless some $0^R \leqslant x$ (false) or $0 \leqslant -1$ (false). So in fact $x = 0$.

More generally, the argument proves that if every $x^L < 0$ and every $x^R > 0$, then $x = 0$, so for instance $\{-1, -\frac{1}{2} \mid 2, 3\} = 0$.

But when we have defined $2\frac{1}{2}$ and 17 we shall have to decide about $\{2\frac{1}{2} \mid 17\}$. A first guess might be their mean, $9\frac{3}{4}$, but since we have just seen that the mean rule does not always hold, this seems unlikely. A clue is given in the form of the preceding argument—since we must ask the questions "is $x = y$?" for the various possible y in order of simplicity, the answer should be the *simplest* y that is not prohibited. This rule will be established in Chapter 2, and it implies, for instance, that $\{2\frac{1}{2} \mid 17\} = 3$, and $\{\frac{1}{4} \mid 1\} = \frac{1}{2}$.

The numbers $\frac{1}{4}$, $\frac{3}{4}$, $1\frac{1}{2}$, 3, and so on

Once we have settled all the trivialities like $x \geqslant x$ for all x (which we have begun to take for granted), we can proceed a little faster. For instance, if L and R are sets of numbers chosen from those we already have, then since we suspect these numbers are totally ordered, in any expression $x = \{x^L \mid x^R\}$ we need only consider the greatest x^L (if any) and the least x^R (ditto). This gives us for the next "day" only the numbers

$$0 < \{0 \mid \tfrac{1}{2}\} < \tfrac{1}{2} < \{\tfrac{1}{2} \mid 1\} < 1 < \{1 \mid 2\} < 2 < \{2 \mid \}$$

and their negatives. What are the proper names for these numbers? We suspect that $\{2 \mid \} = 3$, and indeed we can verify that

$$1 + 1 + 1 = \{0 + 1 + 1, 1 + 0 + 1, 1 + 1 + 0 \mid \} = \{2 \mid \}.$$

The equation $\{1 \mid 2\} = 1\frac{1}{2}$ is almost as easy to guess and verify. So we shall make $1\frac{1}{2}$ a permanent name for this number.

The two likely guesses for $\{0 \mid \frac{1}{2}\}$ are $\frac{1}{3}$ and $\frac{1}{4}$. If anything, the first might seem the better guess, since otherwise it's hard to see what $\frac{1}{3}$ will be. But in fact it turns out that $\{0 \mid \frac{1}{2}\}$ is $\frac{1}{4}$—at least we can verify that twice this number is $\frac{1}{2}$. In a similar way, $\{\frac{1}{2} \mid 1\}$ turns out to be $\frac{3}{4}$ rather than $\frac{2}{3}$.

It is now easy to guess the pattern for the numbers which take only finitely much work to define. Let us imagine the numbers created on successive "days", in such a way that on day number n we create all numbers $x = \{L \mid R\}$ for which every member of each of the two sets L, R has already been constructed. We number the day on which 0 was created with the number 0 itself, so that our creation story begins (or began?) on *the zeroth day*.

Then the numbers seem to form a tree, as shown in Fig. 0. Each node of the tree has two "children", namely the first later numbers born just to the left

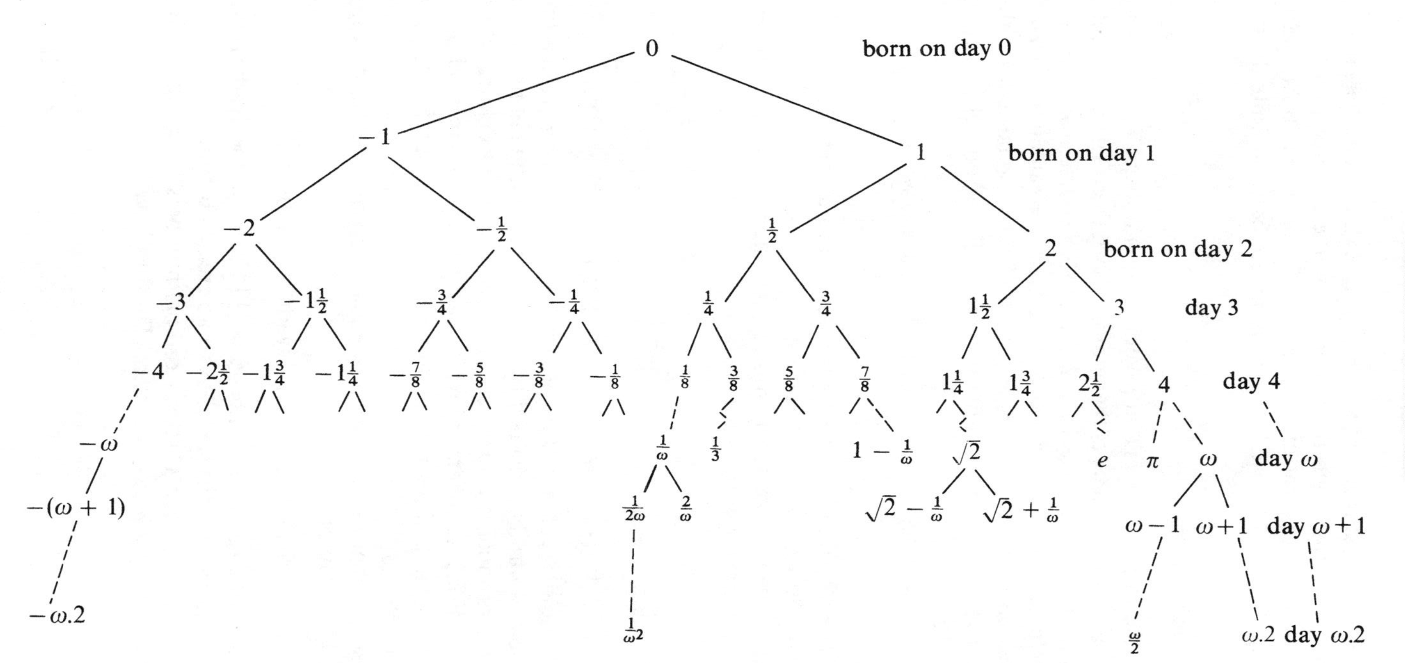

FIG. 0. When the first few numbers were born.

and right of it. We guess that on the n'th day the extreme numbers to be born are n and $-n$, and that each other number is the arithmetic mean of the numbers to the left and right of it. Happily, of course, this turns out to be the case. Supposing all this, we know all numbers born on finite days.

The numbers born on day ω

Of course the process doesn't stop with these numbers. The next day we call day ω. Let's consider some of the numbers born then. The largest number is the number ω itself, defined as $\{0, 1, 2, 3, \ldots \mid \}$. Of course, ω has many other forms, for instance $\omega = \{1, 2, 4, 8, 16, \ldots \mid \}$, or even $\omega = \{\text{all numbers } (m/2^n) \mid \}$. But since the collection of numbers to the left of ω has no largest member in these expressions, we cannot simply eliminate all but one of the numbers appearing on the left.

Of course the most negative number born on day ω will be

$$-\omega = \{ \mid 0, -1, -2, -3, \ldots\}.$$

The smallest positive number born on this day is the number $\{0 \mid 1, \frac{1}{2}, \frac{1}{4}, \frac{1}{8}, \ldots\}$, which turns out to be $1/\omega$, surprisingly and fortunately.

But besides these strange new numbers, some quite ordinary numbers are born at the same time. For instance, we have

$$\tfrac{1}{4} < \tfrac{1}{4} + \tfrac{1}{16} < \tfrac{1}{4} + \tfrac{1}{16} + \tfrac{1}{64} < \ldots < \tfrac{1}{3} < \ldots < \tfrac{1}{2} - \tfrac{1}{8} < \tfrac{1}{2},$$

so we might expect the number

$$\{\tfrac{1}{4}, \tfrac{1}{4} + \tfrac{1}{16}, \tfrac{1}{4} + \tfrac{1}{16} + \tfrac{1}{64}, \ldots \mid \tfrac{1}{2}, \tfrac{1}{2} - \tfrac{1}{8}, \ldots\} = x, \text{ say}$$

to be $\frac{1}{3}$, and behold, it can in fact be proved that $x + x + x = 1$! In a similar way, all of the real numbers defined by Dedekind, including in particular all the remaining rational numbers can be defined as "Dedekind sections" of the dyadic rational numbers (by which we mean the numbers of the form $m/2^n$, m and n integers), rather than as sections of *all* rationals. So $\sqrt{2}$, e, and π are all born on day ω.

It is rather nice that our definition of equality ensures automatically that the number (for example)

$$\{\text{dyadic rationals} < \tfrac{3}{8} \mid \text{dyadic rationals} > \tfrac{3}{8}\}$$

turns out to be the same as the number $\frac{3}{8} = \{\frac{1}{4} \mid \frac{1}{2}\}$, so that the dyadic rationals "recreated" on day ω are "the same" as those created before.

It is also rather nice that Cantor's ordinal numbers (as modified by von Neumann) fit smoothly into our system. Thus we have

$$0 = \{\mid\},\quad 1 = \{0 \mid\},\quad 2 = \{0, 1 \mid\}, \ldots,\quad \omega = \{0, 1, 2, 3, \ldots \mid\},$$

$$\alpha = \{\beta < \alpha \mid\},$$

where von Neumann has

$$0 = \{\},\quad 1 = \{0\},\quad 2 = \{0, 1\}, \ldots,\quad \omega = \{0, 1, 2, \ldots\},\quad \alpha = \{\beta < \alpha\}, \ldots.$$

In other words, the ordinal numbers are those we obtain by requiring always that the set R be empty. We may say that Cantor was only interested in moving ever rightwards, whereas Dedekind stopped to fill in the gaps, so that R was *always* empty for Cantor, *never* empty for Dedekind. It is remarkable that by dropping these restrictions we obtain a theory that is both more general and more easy to work with. (Compare the theory developed in the next chapter with the classical foundation for the real numbers in which we must first construct or postulate the "natural numbers", then rationals as equivalence classes of ordered pairs, then reals as sections of rationals, with negative numbers being introduced at some stage in the process.)

Some more numbers

After ω, the number $\{0, 1, 2, 3, \ldots, \omega \,|\, \} = \omega + 1$ need come as no surprise, but perhaps the number $\{0, 1, 2, 3, \ldots \,|\, \omega\}$ will. This number, call it x, should satisfy $n < x < \omega$ for all finite integers n, in other words, x should be an infinite number less than the "least" infinite number ω. Adding 1 to x, we find the number

$$\{1, 2, 3, \ldots, x \,|\, \omega + 1\} = y, \text{ say.}$$

Here, since $x < \omega$, and $\omega + 1 \not\leq \omega$, we see that $y = \omega$, for the new entries x on the left and $\omega + 1$ on the right have made no difference. So $x + 1 = \omega$, $x = \omega - 1$.

Check that we get the same result on subtracting 1 from ω.

In a similar way, we find successively that

$$\omega - 2 = \{0, 1, 2, 3, \ldots \,|\, \omega, \omega - 1\}, \ldots,$$

$$\omega - n = \{0, 1, 2, 3, \ldots \,|\, \omega, \omega - 1, \omega - 2, \ldots, \omega - (n - 1)\}.$$

Plainly the next number to consider is

$$z = \{0, 1, 2, 3, \ldots \,|\, \omega, \omega - 1, \omega - 2, \ldots\} = \{n \,|\, \omega - n\}, \text{ say.}$$

It should not take the reader too long to verify that $z = \omega/2$. When he has done this, and defined $\omega/4$, $\omega/8, \ldots$ as well, he should be in a position to define $\omega/3$ (for instance), and to verify our assertion that

$$\{0, 1, 2, 3, \ldots \,|\, \omega, \omega/2, \omega/4, \omega/8, \ldots\}$$

is a square root of ω.

Other easy exercises are

$$\left\{0 \,\middle|\, \frac{1}{\omega}\right\} = \frac{1}{2\omega}, \quad \left\{\frac{1}{\omega} \,\middle|\, 1, \tfrac{1}{2}, \tfrac{1}{4}, \ldots\right\} = \frac{2}{\omega}, \quad \left\{0 \,\middle|\, \frac{1}{\omega}, \frac{1}{2\omega}, \frac{1}{4\omega}, \ldots\right\} = \frac{1}{\omega^2},$$

and so on.

If the reader prefers to try his hand at "constructing" new numbers rather than examining values of those given here, let him try to find definitions for $\sqrt[3]{\omega}$, $\omega^{1/\omega}$, $\omega + \pi$, $(\omega + 1)^{-1}$, $\sqrt{(\omega - 1)}$, and to show, making any reasonable assumptions, that they have the properties we should expect.

In the next chapter, we shall prove that the Class of all numbers really is a Field, making no use of any of the supposed "facts" from this chapter. It will be some time before we see so many particular numbers mentioned again. In the third chapter, we shall produce a "canonical form" for numbers, and learn how to manipulate them a little more freely, and in the process will see exactly how general our class of numbers turns out to be.

CHAPTER 1

The Class No is a Field

Ah! why, ye Gods, should two and two make four?
Alexander Pope, "The Dunciad"

PRELIMINARY COMMENTS

There are two problems that arise in the precise treatment which need special comment. The first is that it is necessary to have an expression $\{L \mid R\}$ *existing* even before we have proved that it is a number. The second concerns the fact that equality is a defined relation, which must initially be distinguished from identity.

Games. The construction for numbers generalises immediately to the following construction for what we call *games*.

Construction. If L and R are any two sets of games, then there is a game $\{L \mid R\}$. All games are constructed in this way.

Although games are properly the subject of the first part of this book (where the name will be justified), it is logically necessary to introduce them before numbers. Order-relations and arithmetic operations on games are defined by the same definitions as for numbers. The most important distinction between numbers and general games is that numbers are totally ordered, but games are not—there exist games x and y for which we have neither of $x \geqslant y, y \geqslant x$.

To show that a game $x = \{x^L \mid x^R\}$ is a number, we must show *firstly* that all of the games x^L, x^R are numbers, and *secondly*, that there is no inequality of the form $x^L \geqslant x^R$.

IDENTITY AND EQUALITY

We shall call games x and y *identical* ($x \equiv y$) if their left and right sets are identical—that is, if every x^L is identical to some y^L, every x^R identical to some y^R, and vice versa. Recall that x and y are defined to be *equal* ($x = y$) if and only if we have both $x \geqslant y$ and $y \geqslant x$. The distinction causes no great problems until we come to multiplication, where the trouble is that there can exist equal games x and y for which xz and yz are unequal. But all goes well as long as we restrict ourselves to the multiplication of numbers.

Finally, we note that almost all our proofs are inductive, so that, for instance, in proving something about the pair (x, y) we can suppose that thing already known about all pairs (x^L, y), (x^R, y), (x, y^L), (x, y^R). After a time we feel free to suppress all references to these inductive hypotheses. We remind the reader again that since ultimately we are reduced to questions about members of the empty set, no one of our inductions will require a "basis". The games x^L, x^R will be called the Left, Right *options* of x.

PROPERTIES OF ORDER AND EQUALITY

Recall that $x \geqslant y$ if we have no inequality of form $x^R \leqslant y$ or $x \leqslant y^L$.

THEOREM 0. *For all games x we have*

(i) $x \ngeqslant x^R$,
(ii) $x^L \ngeqslant x$,
(iii) $x \geqslant x$,
(iv) $x = x$.

Proof. (i) Taking y as x^R in the definition of $\geqslant$, and using the inductively true relation $x^R \leqslant x^R$, we see that we cannot have $x \geqslant y$.

(ii) is similar.

(iii) Taking y as x, we now know that we have no $x^R \leqslant y$ and $x \leqslant$ no y^L, whence $x \geqslant y$.

(iv) from $x \geqslant x$ and $x \leqslant x$, we deduce $x = x$.

THEOREM 1. *If $x \geqslant y$ and $y \geqslant z$, then $x \geqslant z$.*

Proof. Since $x \geqslant y$, we cannot have $x^R \leqslant y$, and so by induction we cannot have $x^R \leqslant z$. Similarly we cannot have $x \leqslant z^L$, and so we must have $x \geqslant z$.

Summary. We now know that $\geqslant$ is a partial order relation on games, and that $=$ has the right properties (for instance $x = y$ and $x < z$ imply $y < z$).

THEOREM 2. *For any number x we have $x^L < x < x^R$ for all x^L, x^R. Also, for any two numbers x and y we must have $x \leqslant y$ or $x \geqslant y$.*

Proof. (i) Since we know $x \ngeqslant x^R$, it suffices to prove $x^R \geqslant x$. This will be true unless some $x^{RR} \leqslant x$ or $x^R \leqslant$ some x^L. But the former inductively implies $x^R < x^{RR} \leqslant x$, a contradiction, and the latter is prohibited by the definition of number.

(ii) The inequality $x \ngeqslant y$ implies either some $x^R \leqslant y$ or $x \leqslant$ some y^L, whence either $x < x^R \leqslant y$ or $x \leqslant y^L < y$.

Summary. Numbers are totally ordered.

PROPERTIES OF ADDITION

Definition. $0 = \{\,|\,\}$.
We recall that $x + y = \{x^L + y, x + y^L \,|\, x^R + y, x + y^R\}$.

THEOREM 3. *For all x, y, z we have*

$$x + 0 \equiv x, \qquad x + y \equiv y + x, \qquad (x + y) + z \equiv x + (y + z).$$

Proof.

$$x + 0 \equiv \{x^L + 0 \,|\, x^R + 0\} \equiv \{x^L \,|\, x^R\} \equiv x$$

$$x + y \equiv \{x^L + y, x + y^L \,|\, x^R + y, x + y^R\} \equiv$$
$$\equiv \{y + x^L, y^L + x \,|\, y + x^R, y^R + x\} \equiv y + x.$$

$$(x + y) + z \equiv \{(x + y)^L + z, (x + y) + z^L \,|\, \ldots\} \equiv$$
$$\equiv \{(x^L + y) + z, (x + y^L) + z, (x + y) + z^L \,|\, \ldots\} \equiv$$
$$\equiv \{x^L + (y + z), x + (y^L + z), x + (y + z^L) \,|\, \ldots\} \equiv$$
$$\equiv \ldots \equiv x + (y + z).$$

In each case the middle identity follows from the inductive hypothesis. Proofs like these we call *1-line proofs* even when as here the "line" is too long for our page. We shall meet still longer 1-line proofs later on, but they do not get harder—one simply transforms the left-hand side through the definitions and inductive hypotheses until one gets the right hand side.

Summary. Addition is a commutative Semigroup operation with 0 as zero, even when we demand identity rather than equality.

PROPERTIES OF NEGATION

Recall the definition $-x = \{-x^R \,|\, -x^L\}$.

THEOREM 4. (i) $-(x + y) \equiv -x + -y$
(ii) $-(-x) \equiv x$
(iii) $x + -x = 0$

Proof. (i) and (ii) have easy 1-line proofs. Note that (iii) is an equality rather than an identity. If, say, $x + -x \not\geqslant 0$, we should have some $(x + -x)^R \leqslant 0$, that is, $x^R + -x \leqslant 0$ or $x + -x^L \leqslant 0$. But these are false, since we have by induction $x^R + -x^R \geqslant 0$, $x^L + -x^L \geqslant 0$.

Summary. With equality rather than identity, addition is a commutative Group operation, with 0 for zero, and $-x$ for the negative of x. All this is true for general games.

PROPERTIES OF ADDITION AND ORDER

THEOREM 5. *We have* $y \geqslant z$ *iff* $x + y \geqslant x + z$.

Proof. If $x + y \geqslant x + z$, we cannot have

$$x + y^R \leqslant x + z \text{ or } x + y \leqslant x + z^L,$$

and so by induction we cannot have $y^R \leqslant z$ or $y \leqslant z^L$, so that $y \geqslant z$.

Now supposing $x + y \not\geqslant x + z$, we must have one of

$$x^R + y \leqslant x + z, \quad x + y^R \leqslant x + z, \quad x + y \leqslant x^L + z, \quad x + y \leqslant x + z^L,$$

and if we further suppose $y \geqslant z$, we deduce one of

$$x^R + y \leqslant x + y, \quad x + y^R \leqslant x + y, \quad x + z \leqslant x^L + z, \quad x + z \leqslant x + z^L,$$

all of which imply contradictions by cancellation.

Theorem 5 implies in particular that we have $y = z$ iff $x + y = x + z$, justifying replacement by equals in addition.

THEOREM 6. (i) 0 *is a number*,

(ii) *if* x *is a number, so is* $-x$,

(iii) *if* x *and* y *are numbers, so is* $x + y$.

Proofs. (i) we cannot have $0^L \geqslant 0^R$, since there exists neither a 0^L nor a 0^R.

(ii) From $x^L < x < x^R$ and x^L, x^R numbers, we inductively deduce $-x^R < -x < -x^L$ and $-x^R$, $-x^L$ numbers.

(iii) We deduce inductively that each of

$$x^L + y, x + y^L < x + y < \text{each of } x^R + y, x + y^R,$$

all of $x^L + y$, etc., being numbers.

Summary. Numbers form a totally ordered Group under addition.

PROPERTIES OF MULTIPLICATION

Definition. $1 = \{0 \mid \}$

We recall the definition of multiplication

$$xy = \{x^Ly + xy^L - x^Ly^L,\quad x^Ry + xy^R - x^Ry^R \mid \mid x^Ly + xy^R - x^Ly^R, x^Ry + xy^L - x^Ry^L\}.$$

THEOREM 7. *For all x, y, z we have the identities*

$$x0 \equiv 0,\quad x1 \equiv x,\quad xy \equiv yx,\quad (-x)y \equiv x(-y) \equiv -xy,$$

and the equalities

$$(x + y)z = xz + yz,\quad (xy)z = x(yz).$$

Proof. The identities have easy 1-line proofs. The equalities also have 1-line proofs, as follows:

$$\begin{aligned}(x + y)z &\equiv \{(x + y)^Lz + (x + y)z^L - (x + y)^Lz^L, \ldots \mid \ldots\} \equiv \\ &\equiv \{(x^L + y)z + (x + y)z^L - (x^L + y)z^L, \\ &\quad (x + y^L)z + (x + y)z^L - (x + y^L)z^L, \ldots \mid \ldots\} = \\ &= \{(x^Lz + xz^L - x^Lz^L) + yz,\quad xz + (y^Lz + yz^L - y^Lz^L), \ldots \mid \ldots\} \\ &\equiv xz + yz.\end{aligned}$$

[This fails to yield an identity since the law $x + -x = 0$ is invoked.] The central expression for xyz has four expressions like

$$x^Lyz + xy^Lz + xyz^L - x^Ly^Lz - x^Lyz^L - xy^Lz^L + x^Ly^Lz^L$$

(with perhaps some even number of x^L, y^L, z^L replaced by x^R, y^R, z^R) on the left, and four similar expressions (with an odd number of such replacements) on the right.

Note. We now have the more illuminating form

$$\{xy - (x - x^L)(y - y^L),\quad xy - (x^R - x)(y^R - y) \mid \mid xy + (x - x^L)(y^R - y),\quad xy + (x^R - x)(y - y^L)\}$$

for the product xy.

THEOREM 8. (i) *If x and y are numbers, so is xy*

(ii) *If $x_1 = x_2$, then $x_1y = x_2y$*

(iii) *If $x_1 \leqslant x_2$, and $y_1 \leqslant y_2$, then $x_1y_2 + x_2y_1 \leqslant x_1y_1 + x_2y_2$, the conclusion being strict if both the premises are.*

Proof. We shall refer to the inequality of (iii) as $P(x_1, x_2 : y_1, y_2)$. Note that if $x_1 \leqslant x_2 \leqslant x_3$, then we can deduce $P(x_1, x_3 : y_1, y_2)$ from the inequalities $P(x_1, x_2 : y_1, y_2)$ and $P(x_2, x_3 : y_1, y_2)$ by adding these and cancelling common terms from the two sides.

Now to prove (i), we observe first that inductively, all options of xy are numbers, so that we have only to prove a number of inequalities like

$$x^{L_1}y + xy^L - x^{L_1}y^L < x^{L_2}y + xy^R - x^{L_2}y^R.$$

But if $x^{L_1} \leqslant x^{L_2}$ we have

$$x^{L_1}y + xy^L - x^{L_1}y^L \leqslant x^{L_2}y + xy^L - x^{L_2}y^L < x^{L_2}y + xy^R - x^{L_2}y^R$$

(these two inequalities reducing respectively to $P(x^{L_1}, x^{L_2} : y^L, y)$ and $P(x^{L_2}, x : y^L, y^R)$), while if $x^{L_2} \leqslant x^{L_1}$ we have instead

$$x^{L_1}y + xy^L - x^{L_1}y^L < x^{L_1}y + xy^R - x^{L_1}y^R \leqslant x^{L_2}y + xy^R - x^{L_2}y^R.$$

(these being $P(x^{L_1}, x : y^L, y^R)$ and $P(x^{L_2}, x^{L_1} : y, y^R)$).

Now to prove (ii). This implication follows immediately from the fact that every Left option of either is strictly less than the other, and every Right option strictly greater, the relevant inequalities all being easy.

If $x_1 = x_2$ or $y_1 = y_2$ we can use (ii) to show that the terms on the Left of (iii) are equal to those on the Right.

So we need only consider the case $x_1 < x_2$, $y_1 < y_2$. Since $x_1 < x_2$, we have either $x_1 < x_1^R \leqslant x_2$ or $x_1 \leqslant x_2^L < x_2$, say the former. But then $P(x_1, x_2 : y_1, y_2)$ can be deduced from $P(x_1, x_1^R : y_1, y_2)$ and $P(x_1^R, x_2 : y_1, y_2)$, of which the latter is strictly simpler than the original. A similar argument now reduces our problem to proving strict inequalities of the four forms

$$P(x^L, x : y^L, y), \quad P(x^L, y : y, y^R), \quad P(x, x^R : y^L, y), \quad \text{and} \quad P(x, x^R : y, y^R)$$

which merely assert that xy has the right order relations with its options.

THEOREM 9. *If x and y are positive numbers, so is xy.*

Proof. This follows from $P(0, x : 0, y)$.

Summary. Numbers form a totally ordered Ring. Note that in view of Theorem 8 and the distributive law, we can assert, for example, that $x \geqslant 0$, $y \geqslant z$ together imply $xy \geqslant xz$, and that if $x \neq 0$, we can deduce $y = z$ from $xy = xz$.

PROPERTIES OF DIVISION

We have just shown that if there is any number y such that $xy = t$, then y is uniquely determined by x and t provided that $x \neq 0$. We must now show how to produce such a y. It suffices to show that for positive x there is a number y such that $xy = 1$. We first put x into a sort of standard form.

LEMMA. *Each positive x has a form in which 0 is one of the x^L, and every other x^L is positive.*

Proof. Let y be obtained from x by inserting 0 as a new Left option, deleting all negative Left options. Then it is easy to check that y is a number, and that $y = x$.

We write $x = \{0, x^L \mid x^R\}$ in this section, and restrict use of the symbol x^L to the positive Left options of x. (Note that all the x^R are automatically positive.)

Now we shall define a number y, explain the definition, and prove that y is a number and that $xy = 1$.

Definition

$$y = \left\{0, \frac{1 + (x^R - x)y^L}{x^R}, \frac{1 + (x^L - x)y^R}{x^L} \,\middle|\, \frac{1 + (x^L - x)y^L}{x^L}, \frac{1 + (x^R - x)y^R}{x^R}\right\}$$

Note that expressions involving y^L and y^R appear in the definition of y. It is this that requires us to "explain" the definition. The explanation is that we regard these parts of the definition as defining new options for y in terms of old ones. So even the definition of this y is an inductive one.† [This is in addition to the "other" induction by which we suppose that inverses for the x^L and x^R have already been found.]

THEOREM 10. *We have* (i) $xy^L < 1 < xy^R$ *for all* y^L, y^R.

(ii) *y is a number.*

(iii) $(xy)^L < 1 < (xy)^R$ *for all* $(xy)^L, (xy)^R$.

(iv) $xy = 1$.

Proof. We observe that the options of y are defined by formulae of the form

$$y'' = \frac{1 + (x' - x)y'}{x'}$$

where y' is an "earlier" option of y, and x' some non-zero option of x. This formula can be written

$$1 - xy'' = (1 - xy')\frac{x' - x}{x'}$$

which shows that y'' satisfies (i) if y' does. Plainly 0 does. Part (ii) now follows,

† To see how the definition works, take $x = \{0, 2 \mid \} = 3$. Then there is no x^R and the only x^L is 2, so $x^L - x = -1$ and the formula for y becomes $y = \{0, \frac{1}{2}(1 - y^R) \mid \frac{1}{2}(1 - y^L)\}$. The initial value $y^L = 0$ gives us $\frac{1}{2}(1 - 0) = \frac{1}{2}$ for a new y^R, whence $\frac{1}{2}(1 - \frac{1}{2}) = \frac{1}{4}$ as a y^L, then $\frac{1}{2}(1 - \frac{1}{4}) = \frac{3}{8}$ for a y^R, and so on, yielding $y = \{0, \frac{1}{4}, \frac{5}{16}, \ldots \mid \frac{1}{2}, \frac{3}{8}, \ldots\}$, which certainly looks like $\frac{1}{3}$.

since we cannot have any inequality $y^L \geqslant y^R$. The typical form of an option of xy is $x'y + xy' - x'y'$, which can be written as $1 + x'(y - y'')$ with the above definition of y'', and this suffices to prove (iii). For (iv), we observe first that $z = xy$ has a left option 0 (take $x^L = y^L = 0$), and that (iii) asserts that $z^L < 1 < z^R$ for all z^L, z^R. Then

$z \geqslant 1$, since no $z^R \leqslant 1$, and $z \leqslant$ no 1^L (since some $z^L = 0$), and also

$1 \geqslant z$, since no $1^R \leqslant z$, and $1 \leqslant$ no z^L,

so that indeed $z = 1$.

Summary. The Class **No** of all numbers forms a totally ordered Field.

Clive Bach has found a similar definition for the square root of a non-negative number x. He defines

$$\sqrt{x} = y = \left\{ \sqrt{x^L}, \frac{x + y^L y^R}{y^L + y^R} \,\middle|\, \sqrt{x^R}, \frac{x + y^L y^{L*}}{y^L + y^{L*}}, \frac{x + y^R y^{R*}}{y^R + y^{R*}} \right\}$$

where x^L and x^R are non-negative options of x, and y^L, y^{L*}, y^R, y^{R*} are options of y chosen so that no one of the three denominators is zero. We shall leave to the reader the easy inductive proof that this is correct.

Martin Kruskal has pointed out that the options of $1/x$ can be written in the form

$$\frac{1 - \Pi\left(1 - \dfrac{x}{x_i}\right)}{x}$$

where the denominator x cancels formally, the x_i denote positive options of x, and the product may be empty. This is a Left option of $1/x$ just when an *even* number of the x_i are Left options of x. There is a similar closed form for Bach's definition of $\sqrt{x}$.

CHAPTER 2

The Real and Ordinal Numbers

Don't let us make imaginary evils, when you know we have so many real ones to encounter.

Oliver Goldsmith, "The Good-Natured Man"

The following theorem gives us a very easy way of evaluating particular numbers. We call it *the simplicity theorem.*

THEOREM 11. *Suppose for* $x = \{x^L | x^R\}$ *that some number* z *satisfies* $x^L \not\geqslant z \not\geqslant x^R$ *for all* x^L, x^R, *but that no option of* z *satisfies the same condition. Then* $x = z$.

[Note: this holds even when x is only given to be a game.]

Proof. We have

$$x \geqslant z \text{ unless some } x^R \leqslant z \text{ (no!) or } x \leqslant \text{ some } z^L.$$

But from $x \leqslant z^L$, we can deduce $x^L \not\geqslant x \leqslant z^L < z \not\geqslant x^R$ for all x^L, x^R, from which we have $x^L \not\geqslant z^L \not\geqslant x^R$, contradicting the supposition about z. So $x \geqslant z$, similarly $z \geqslant x$, and so $x = z$.

The main assertion of the theorem is that when x is given as a number, it is always the *simplest* number lying between the x^L and the x^R, where *simplest* means *earliest created.* [For if z is this simplest number, the simpler numbers z^L, z^R cannot satisfy the same condition.] But the exact version presented above has several advantages, since it holds when x is given as a game not necessarily known to equal a number, and it is perhaps not quite obvious exactly what is meant by "the simplest number such that...". In the applications below, there is never any problem.

THEOREM 12. *If* x *is a rational number whose denominator divides* 2^n, *then* $x = \{x - (1/2^n) | x + (1/2^n)\}$.

Proof. For $n = 0$ the theorem holds, since it asserts that x is the simplest number between $x - 1$ and $x + 1$, whereas we know that in fact it is, if positive, the simplest number greater than $x - 1$, if negative the simplest number less than $x + 1$, and if zero the simplest number of all. [These statements follows from the usual definition of integers as sums of 1 or -1.]

For $n > 0$, we double $z = \{x - (1/2^n) \mid x + (1/2^n)\}$ to see that $2z$ is the simplest number between $z + x - (1/2^n)$ and $z + x + (1/2^n)$. Since z certainly lies between $x - (1/2^n)$ and $x + (1/2^n)$ these limits are between $2x - (1/2^{n-1})$ and $2x + (1/2^{n-1})$, and by induction $2x$ is the simplest number between these limits, so that $2z = 2x$, $z = x$.

Theorem 12 justifies all the assertions of Chapter 0 about numbers born on finite days. Every such number is a *dyadic rational* number, that is, a rational number of the form $m/2^n$. Of course, we can speak of "the" rational number p/q without ambiguity, since we have shown that **No** is a totally ordered Field, and therefore contains a uniquely defined image of each rational number, supposed defined in any of the usual ways.

CONTAINMENT OF THE REAL NUMBERS

Definition. x is a *real number* if and only if $-n < x < n$ for some integer n, and

$$x = \{x - 1, x - \tfrac{1}{2}, x - \tfrac{1}{3}, \ldots \mid x + 1, x + \tfrac{1}{2}, x + \tfrac{1}{3}, \ldots\},$$

or in short, $x = \{x - (1/n) \mid x + (1/n)\}_{n>0}$. [It is to be understood that n ranges over the positive integers.]

THEOREM 13. (i) *Dyadic rationals are real numbers.*

(ii) $-x$, $x + y$, *and* xy *are real if* x *and* y *are.*

(iii) *Each real number has a unique expression in the form* $\{L \mid R\}$, *where* L *and* R *are non-empty sets of rationals,* L *has no greatest,* R *no least, and there is at most one rational in neither* L *nor* R. *Also,* $y' < y \in L$ *implies* $y' \in L$, $z' > z \in R$ *implies* $z' \in R$.

(iv) *Each section* $\{L \mid R\}$ *as in* (iii) *equals a unique real number.*

Proof. (i) follows from Theorems 11 and 12. (ii) follows from the formulae defining the operations (it might be helpful to use the version of the product formula in the note before Theorem 8). As for (iii), for any real number x, let L = the set of rationals less than x, R = the set of rationals greater than x. Then L and R are non-empty by the condition $-n < x < n$ for some n. Also every member of L is less than $x - (1/n)$ for some n, and so we can add $1/2n$ and still be less than x. This shows that L has no greatest, and similarly R no least, member. A rational in neither L nor R must equal x, so at most one is in neither. Since the expression is obviously unique, this proves (iii).

As for (iv), note that $\{L \mid R\}$ is certainly *some* number, x, say, and that easily $-n < x < n$ for some integer n. So we need only show

$$x = \left\{x - \frac{1}{n} \middle| x + \frac{1}{n}\right\}_{n>0}.$$

But since L has no greatest, for any $y \in L$ we have $y + (1/n) \in L$ for all sufficiently large n. This shows that for sufficiently large n there is a member of L greater than $x - (1/n)$ and similarly a member of R less than $x + (1/n)$, which suffices.

Note. We could obviously replace rationals throughout by dyadic rationals in (iii) and (iv). On doing so, we deduce that every real number not a dyadic rational is born on day ω, as asserted in Chapter 0.

Summary. The real numbers as defined here behave exactly like the real numbers defined in any of the more usual ways. So we shall use the name $\mathbb{R}$ for the set of all real numbers.

THE LOGICAL THEORY OF REAL NUMBERS

We have here regarded the ordinary real numbers and their theory as known, so that Theorem 13 merely serves to identify "our" real numbers with the familiar ones. But of course one could use our ideas to give a new logical foundation for the real numbers. We digress to discuss the usual classical treatments and the advantages and disadvantages of the possible new approach.

Figure 1 shows the lattice of inclusions between the sets $\mathbb{Z}$, $\mathbb{Q}$, $\mathbb{R}$ of integers, rationals, and reals, and the corresponding sets $\mathbb{Z}^+$, $\mathbb{Q}^+$, $\mathbb{R}^+$ of positive

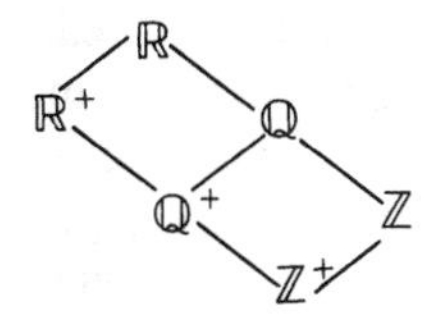

FIG. 1.

elements. [It does not matter very much whether we add here the number 0 or not.] We shall suppose $\mathbb{Z}^+$ and its properties already known. Then one sees at once that there are several possible paths through the lattice from $\mathbb{Z}^+$

to $\mathbb{R}$. Some experience in teaching convinces one that there is a unique best possible path, which is *not* one that seems natural at first sight!

For $\mathbb{X} = \mathbb{Z}$ or $\mathbb{Q}$ or $\mathbb{R}$ we can proceed from $\mathbb{X}^+$ to $\mathbb{X}$ by introducing ordered pairs (a, b) meaning $a - b$, and the equivalence relation $(a, b) \sim (c, d)$ iff $a + d = b + c$. [The alternative of adding new elements 0 and $-x$ $(x \in \mathbb{X}^+)$ leads to too much case-splitting.]

Similarly we can proceed from $\mathbb{Z}$ to $\mathbb{Q}$ or $\mathbb{Z}^+$ to $\mathbb{Q}^+$ by introducing ordered pairs (a, b) meaning a/b and the equivalence relation $(a, b) \sim (c, d)$ iff $ad = bc$.

We proceed from $\mathbb{Q}$ to $\mathbb{R}$ or $\mathbb{Q}^+$ to $\mathbb{R}^+$ by the method of Dedekind sections, or that of Cauchy sequences.

In practice the main problem is to avoid tedious case discussions. [Nobody can seriously pretend that he has ever discussed even eight cases in such a theorem—yet I have seen a presentation in which one theorem actually had 64 cases!] Now if we define $\mathbb{R}$ in terms of Dedekind sections in $\mathbb{Q}$, then there are at least four cases in the definition of the product xy according to the signs of x and y. [And zero often requires special treatment!] This entails eight cases in the associative law $(xy)z = x(yz)$ and strictly more in the distributive law $(x + y)z = xz + yz$ (since we must consider the sign of $x + y$). Of course an elegant treatment will manage to discuss several cases at once, but one has to work very hard to find such a treatment.

This discussion convinces *me* that if one is to use Dedekind sections then the best treatment does not use the branch of our lattice from $\mathbb{Q}$ to $\mathbb{R}$, and so must be the unique shortest path passing through $\mathbb{R}^+$. This seems surprising, since the algebraic theory (introduction of negatives and inverses) should naturally be logically prior to the analytic (limits, etc.).

[The reader should be cautioned about difficulties in regarding the construction of the reals as a particular case of the completion of a metric space. If we take this line, we plainly must not start by defining a metric space as one with a real-valued metric! So initially we must allow only rational values for the metric. But then we are faced with the problem that the metric on the completion must be allowed to have arbitrary real values!

Of course, the problem here is not actually insoluble, the answer being that the completion of a space whose metric takes values in a field $\mathbb{F}$ is one whose metric takes values in the completion of $\mathbb{F}$. But there are still sufficient problems in making this approach coherent to make one feel that it is simpler to first produce $\mathbb{R}$ from $\mathbb{Q}$, and later repeat the argument when one comes to complete an arbitrary metric space, and of course this destroys the economy of the approach. My own feeling is that in any case the apparatus of Cauchy sequences is logically too complicated for the simple passage from $\mathbb{Q}$ to $\mathbb{R}$—one should surely wait until one has the real numbers before doing a piece of analysis!]

This discussion should convince the reader that the construction of the real numbers by any of the standard methods is really quite complicated. Of course the main advantage of an approach like that of the present work is that there is just one kind of number, so that one does not spend large amounts of time proving the associative law in several different guises. I think that this makes it the simplest so far, from a purely logical point of view.

Nevertheless there are certain disadvantages. One that can be dealt with quickly is that it is quite tricky to make the process *stop* after constructing the reals! We can cure this by adding to the construction the proviso that if L is non-empty but with no greatest member, then R is non-empty with no least member, and vice versa. This happily restricts us exactly to the reals.

The remaining disadvantages are that the dyadic rationals receive a curiously special treatment, and that the inductive definitions are of an unusual character. From a purely logical point of view these are unimportant quibbles (we discuss the induction problems later in more detail), but they would predispose me against teaching this to undergraduates as "the" theory of real numbers.

There is another way out. If we adopt a classical approach as far as the rationals $\mathbb{Q}$, and then define the reals as sections of $\mathbb{Q}$ with the definitions of addition and multiplication given in this book, then all the formal laws have 1-line proofs and there is no case-splitting. The definition of multiplication seems complicated, but is fairly easy to motivate. Altogether, this seems the easiest possible approach.

[Perhaps I may add some comments about the multiplication definition. In fact the whole theory was developed even as far as a version of the canonical form theorem of Chapter 3 before any general notion of product appeared, and at first the product was defined in terms of canonical forms. Only several weeks' hard thought, sustained by the conviction that there *must* be a "genetic" definition, finally led to the "correct" formula. The genetic definition of $1/x$ at the end of Chapter 1 only appeared a year later.]

CONTAINMENT OF THE ORDINAL NUMBERS

Definition. α is an *ordinal number* if α has an expression of the form $\alpha = \{L \mid \}$. [*Note that α is then automatically a number.*]

THEOREM 14. *For any x, the class of all ordinal numbers $\not\geqslant x$ is a set* (*i.e. not a proper Class*).

Proof. Since there is no α^R, the condition $\alpha \not\geqslant x$ implies $\alpha \leqslant$ some x^L, and so $\alpha < x^L$ or $\alpha = x^L$. Since the collection of $\alpha <$ any particular x^L is a set by induction, α belongs to a union of a set of sets, and so to a certain set.

THEOREM 15. *For each ordinal α, we have $\alpha = \{\text{ordinals } \beta < \alpha \mid \}$. In any non-empty Class C of ordinals there is a least. For any set S of ordinals there is an ordinal α greater than every member of S.*

Proof. The first part is immediate from the simplicity theorem and the fact that the collection of $\beta < \alpha$ *is* a set. For the second part, we observe that the collection L of all β less than all $\alpha \in C$ is a set, for since C is non-empty L is included in the set of all $\beta <$ *some* $\alpha \in C$. Then defining $\delta = \{L \mid \}$, we find that for all $\alpha \in C$ we have $\alpha \geqslant \delta$, since there is no α^R, and we never have $\alpha \leqslant \delta^L$. Then if $\alpha > \delta$ for all $\alpha \in C$, we get $\delta \in L$, so $\delta < \delta$, a contradiction, and so δ must be equal to some member of C. Finally, the ordinal $\{S \mid \}$ is greater than every member of S.

Summary. We have proved enough to show that there is a one-to-one order-preserving correspondence between the ordinal numbers as defined here and as defined in any of the more usual ways. So we shall use **On** for the Class of all ordinal numbers.

Note. We have regarded the ordinal numbers and their properties as known, so that Theorem 15 merely identifies "our" ordinal numbers with the familiar ones. Naturally it would be possible to develop the logical theory of ordinals directly from our approach. But the standard set theory of Zermelo and Fraenkel does not seem to be the right vehicle in which to develop such a suggestion, since obviously it should be modified so as to allow two notions of membership (Left and Right) first. There is no logical problem, but we prefer to postpone the discussion till later.

The reader should be aware that the operations $\alpha + \beta$ and $\alpha\beta$ as defined here are not the usual *ordinal* operations, but rather the *maximal* sum and product (sometimes called the *natural* sum and product) which can be obtained by treating the Cantor Normal Form like a polynomial. [The *maximal* sum $\alpha + \beta$ is the largest order-type of any well-ordered set $A \cup B$ for which A and B have the respective order-types α and β. The *ordinal* sum is the order-type of such a union in which A precedes B. There are similar definitions of the two product notions.]

We consider a generalization of the Cantor Normal Form in Chapter 3, and in the first part of the book we shall define an operation $G:H$ (for all games G, H) which will generalise the notion of ordinal sum.

CHAPTER 3

The Structure of the General Surreal Number

We admit, in Geometry, not only infinite magnitudes, that is to say, magnitudes greater than any assignable magnitude, but infinite magnitudes infinitely greater, the one than the other. This astonishes our dimension of brains, which is only about six inches long, five broad, and six in depth in the largest heads.

Voltaire, Article "Infinity", in A Philosophical Dictionary, Boston 1881

We return to the tree of numbers sketched in Chapter 0, and make precise some of the notions described there. Greek letters $\alpha, \beta, \gamma, \ldots$ will denote arbitrary *ordinal* numbers.

For each ordinal α we define a set M_α of numbers by setting $x = \{x^L \mid x^R\}$ in M_α if all the x^L and x^R are in the union of all the M_β for $\beta < \alpha$. Then we set $O_\alpha = \bigcup_{\beta<\alpha} M_\beta$, and $N_\alpha = M_\alpha \setminus O_\alpha$. Then in the terminology of Chapter 0 (to which we shall adhere):

M_α is the set of numbers born on or before α (Made numbers),

N_α is the set of numbers born first on day α (New numbers), and

O_α is the set of numbers born before day α (Old numbers).

Now each $x \in N_\alpha$ defines a Dedekind section L, R of O_α, if we set

$$L = \{y \in O_\alpha \mid y < x\}, \quad \text{and} \quad R = \{y \in O_\alpha \mid y > x\}.$$

Since the simplicity theorem tells us that then $x = \{L \mid R\}$, we can regard $M_\alpha = O_\alpha \cup N_\alpha$ as the union of O_α together with all its sections, in the natural order.

Now let $x \in N_\alpha$. Then for each $\beta < \alpha$, x defines a section in O_β, and this section defines a unique point $x_\beta \in N_\beta$. We call x_β the *βth approximation* to x, and extend the definition by writing $x_\beta = x$ for all $\beta \geqslant \alpha$. The reader who glances at Figure 0 will see that the successive approximations to $\sqrt{2}$ are $0, 1, 2, 1\frac{1}{2}, 1\frac{1}{4}, 1\frac{3}{8}, \ldots, \sqrt{2}$, 0 being the 0th approximation and $\sqrt{2}$ the ωth. These numbers are just the nodes above $\sqrt{2}$ on the tree.

THEOREM 16. *Every number x is in a unique set N_α.*
(The ordinal number α is called the *birthday* of x.)

Proof. We suppose this is true of all x^L, x^R. If β is some ordinal greater than the birthdays of all x^L, x^R, then x is certainly in M_β, and so in some N_α, $\alpha \leqslant \beta$.

This theorem assures us that the successive approximations are defined for all numbers x, and they "converge" to x in the sense that they all coincide with x for sufficiently large β.

Now for each $\beta < \alpha$ (the birthday of x) we define a sign s_β (+ or −) as the sign of the number $x - x_\beta$. We extend the definition by writing $s_\beta = 0$ for all $\beta \geqslant \alpha$. In this way, we have assigned to each number x a sequence of signs + or − below some ordinal, 0 beyond, which we call the *sign-expansion* of x.

We now order such sign-sequences lexicographically by the conditions:

$(s) < (t)$ iff for some α we have

$$s_\beta = t_\beta \text{ for all } \beta < \alpha, \text{ but } s_\alpha < t_\alpha,$$

it being understood that $- < 0 < +$.

THEOREM 17. *Let x and y have sign-expansions (s) and (t). Then we have* $x < y$, $x = y$, $x > y$ according as $(s) < (t)$, $(s) = (t)$, $(s) > (t)$.

Proof. If $(s) < (t)$, suppose $s_\beta = t_\beta$ for all $\beta < \alpha$, but $s_\alpha < t_\alpha$. Then $x_\beta = y_\beta$ by induction for all $\beta < \alpha$, but $x_\alpha < y_\alpha$. (It is obvious that the sign-expansion of x_β is simply that of x truncated by making $s_\gamma = 0$ for all $\gamma \geqslant \beta$). The sections defined by x and y in O_α now show that $x < y$. If $(s) = (t)$, we find that x and y define the same section of O_α, where α is their common birthday, and so $x = y$.

THEOREM 18. *For an arbitrary sequence (s) of signs + or − below some ordinal α, 0 beyond, there exists a number x whose sign-expansion is (s).*

Proof. Let $s(\beta)$ denote the expansion truncated at β—that is to say, the sequence (t) defined by $t_\gamma = s_\gamma$ $(\gamma < \beta)$, $t_\gamma = 0$ $(\gamma \geqslant \beta)$. Then by induction, for each $\beta < \alpha$, there exists a number x_β whose sign-expansion is $s(\beta)$. Then we consider the number

$$x = \{x_\beta \text{ for which } s(\beta) < (s) \mid x_\beta \text{ for which } s(\beta) > (s)\}.$$

Plainly the birthday of x is at most α, and s_β is the sign of $x - x_\beta$ for all $\beta < \alpha$, so that x has the desired signs.

Summary. The correspondence between numbers and their sign-expansions is one-to-one and order-preserving.

We regard these results as justifying all assertions made about the tree of numbers considered in Chapter 0, extended to all possible ordinal depths.

Here is a simple rule, due to Elwyn Berlekamp, by which we can read off the value of a *real* number from its sign-expansion. We can suppose that the expansion begins with a $+$, for we change signs of numbers by changing all the signs in their expansions. If the expansion consists just of n $+$ signs, the number is n. Otherwise, bracket the first $-$ with the preceding $+$, when the number of $+$ signs before the bracket defines the integer part, and the signs after the bracket yield the ordinary binary expansion of the fractional part when we read 1 for $+$ and 0 for $-$, adding a final 1 when the expansion is finite.

Thus $++++-+--$ yields $+++(+-)+-- = 3{\cdot}1001 = 3\frac{9}{16}$. In short, the signs before the bracket are interpreted in "unary", those after in binary. The rule can be extended so as to yield the sign-expansions of, for instance, real multiples of ω. Thus since $+-+$ is the expansion of $\frac{3}{4}$, $+^{\omega}\,-^{\omega}\,+^{\omega}$ is that of $\frac{3}{4}\omega$. We shall give later the general rule by which one finds the sign-expansion from the canonical form (see below) or vice versa. Sign-expansions are connected with the generalisation $G{:}H$ of the ordinal sum that appears in the theory of many games, notably the unimpartial form of Hackenbush. The sign-expansion of $x{:}y$ is obtained by following that of x by that of y.

THE ω-MAP

We now define a function ω^x that plays an important role in the theory, and can be thought of as the xth power of ω. More precisely, this is an *ordinal power*, which is not an instance of the "analytic" power operation x^y defined on page 38.

We say that positive numbers x and y are *commensurate* if for some positive integer n we have $x < ny$, $y < nx$. Plainly this is an equivalence relation whose equivalence Classes are *convex* (that is, if $x < z < y$ and x and y are commensurate, then z is commensurate with both). It follows that there is a unique simplest number in each commensurate class, and these numbers we call *leaders*. We obtain the ω-map by letting ω^0 be the simplest leader of all (namely 1), then ω^1 and ω^{-1} be the simplest leaders to the right and left of ω^0 (namely ω and $1/\omega$), and so on. [Thus $\omega^{\frac{3}{4}}$ will be the simplest leader between $\omega^{\frac{1}{2}}$ and ω^1.] The same effect is achieved by the more formal definition

$$\omega^x = \{0, r\omega^{x^L} \mid r\omega^{x^R}\}$$

where r denotes a variable ranging over all positive real numbers. (We shall also use s in this sense.)

THEOREM 19. *Each positive number is commensurate with some* ω^y.

Proof. We can write x in the form $\{0, x^L \mid x^R\}$, where x^L and x^R now denote positive numbers. Each x^L is commensurate with some ω^{y^L} (say) and each x^R with ω^{y^R}. If x is commensurate with one of its options, we are done. If not, we can add all numbers $r\omega^{y^L}$ as Left options and all $r\omega^{y^R}$ as Right options, and we then see that $x = \omega^y$, where y is the number $\{y^L \mid y^R\}$.

THEOREM 20. $\omega^0 = 1$, $\omega^{-x} = 1/\omega^x$, $\omega^{x+y} = \omega^x . \omega^y$.

Proof. The first part is trivial, and the second follows from the first and third. Let $X = \omega^x$, $Y = \omega^y$, and let X' and Y' be the typical options of X and Y. Then the typical option of XY is $X'Y + XY' - X'Y'$. If Y' is 0, this is $X'Y$, and if X' is 0, it is XY'. Otherwise we can suppose $X' = r\omega^{x'}$, $Y' = s\omega^{y'}$, when the formula becomes

$$r\omega^{x'+y} + s\omega^{x+y'} - rs\omega^{x'+y'}$$

by induction.

When this is positive, it lies between two positive real multiples of ω^z, where z is the largest of the three indices, which is always one of $x' + y$ and $x + y'$. We have said enough to show that

$$\omega^x . \omega^y = \{0, r\omega^{x^L+y}, s\omega^{x+y^L} \mid r\omega^{x^R+y}, s\omega^{x+y^R}\} = \omega^{x+y}.$$

Summary. ω^x does indeed behave like the xth power of ω. Those familiar with the normal arithmetic of ordinals will have no difficulty in showing that ω^α is the ordinal usually so called.

THE NORMAL FORM OF x

Let x be an arbitrary positive number, and ω^{y_0} the unique leader commensurate with x. Then we can divide the reals into two classes by putting t into L or R according as $\omega^{y_0} . t \leqslant x$ or $\omega^{y_0} . t > x$. Then L and R are non-empty, since for suitably large n we have $-n \in L$, $n \in R$, and so by the theory of real numbers, one of L and R has an extremal point r_0, say. Write

$$x = \omega^{y_0} . r_0 + x_1.$$

It follows that x_1 is *small compared to* x, that is, that nx_1 is between x and $-x$ for all integers n. If x_1 is not zero, we can produce in a similar way numbers r_1, y_1 such that $x_1 = \omega^{y_1} . r_1 + x_2$, where x_2 is small compared to x_1.

If again x_2 is non-zero, we can continue, producing an expansion

$$x = \omega^{y_0} . r_0 + \omega^{y_1} . r_1 + \ldots + \omega^{y_{n-1}} . r_{n-1} + x_n$$

which will terminate painlessly if any x_n is zero. But usually the expansion will continue for more than ω steps, so that we must say exactly what we mean.

Suppose that for each $\beta <$ some α we have already defined the *β-term* $\omega^{y_\beta} . r_\beta$ of x. Then we shall define the formal sum $\sum_{\beta<\alpha} \omega^{y_\beta} . r_\beta$ to be the *simplest* number whose β-term is $\omega^{y_\beta} . r_\beta$ for all $\beta < \alpha$. Write $x = \sum_{\beta<\alpha} \omega^{y_\beta} . r_\beta + x_\alpha$. Then if x_α is zero we define the α-term of x to be 0, and otherwise $\omega^{y_\alpha} . r_\alpha$, where ω^{y_α} is commensurate with $|x_\alpha|$ and $x_\alpha - \omega^{y_\alpha} . r_\alpha$ is small compared to x_α. This defines the α-term for all ordinals α.

Now for each α the partial sum $\sum_{\beta<\alpha} \omega^{y_\beta} . r_\beta$ is the simplest number having the same β-terms as x for all $\beta < \alpha$, and so all these partial sums must belong to the set M_γ, where γ is the birthday of x. It follows that the partial sums cannot be distinct for all ordinals α, so that the α-term must vanish for some α, and so for all subsequent α. We have therefore proved:

THEOREM 21. *For each x we can define a unique expression $\sum_{\beta<\alpha} \omega^{y_\beta} . r_\beta$, (the* normal form *of x) in which α denotes some ordinal, the numbers r_β $(\beta < \alpha)$ are non-zero reals, and the numbers y_β form a descending sequence of numbers. Normal forms for distinct x are distinct, and every form satisfying these conditions occurs.*

(The last sentence is easy.)

This theorem can be interpreted as showing that the structure of **No** as a Field can be obtained from its structure as an additive Group by means of the Malcev–Neumann transfinite power-series construction. (The Malcev–Neumann construction in general is discussed in P. M. Cohn's "Universal Algebra", p 276.) But the following discussion shows that this remark does not suffice in itself for a *definition* of the arithmetic structure of **No**.

In the next chapter we shall use normal forms to see that the Field **No** is real-closed. In the rest of this one, we shall merely use them to examine some particularly interesting numbers.

IRREDUCIBLE NUMBERS

Can any index in the normal form of α have the same birthday as α? If not then the normal form yields an expression for x in terms of (real and ordinal numbers and) simpler numbers, so that we can call x reducible. Suppose the index y_α in the α-term of x has the same birthday as x. Then it is easy to see that $\omega^{y_\alpha} . r_\alpha$ is the *last* term in the normal form of x, and that $r_\alpha = \pm 1$. [This is because the numbers

$$\sum_{\beta<\alpha} \omega^{y_\beta} . r_\beta \pm \omega^{y_\alpha}$$

are constructed strictly before

$$\sum_{\beta<\alpha} \omega^{y_\beta} . r_\beta + (\omega^{y_\alpha} . r_\alpha + \text{smaller})$$

if the bracketed term here is not $\pm\omega^{y_\alpha}$.]

So in this case, we can write $x = x' \pm \omega^y$, where x' is born before x, ω^y is small compared to x, and y has the same birthday as x. If y is reducible in the sense above, then by inserting the normal form for y we obtain an expression for x in terms of simpler numbers, and so we regard x as reducible in this case also. In the hard cases, we shall find

$$x = x' \pm \omega^y, \quad y = y' \pm \omega^z, \quad z = z' \pm \omega^t, \ldots \text{(to just } \omega \text{ terms).}$$

It is only these numbers which we shall finally call *irreducible.*

The irreducible numbers generalise the concept of ordinal ε-numbers, and it is not hard to see that the birthday of any irreducible number is an ε-number.

CONTINUED EXPONENTIALS FOR IRREDUCIBLES

The *continued exponential* expression for the number x we have just considered is

$$x = x' \pm \omega^{y' \pm \omega^{z' \pm \omega^{t' \pm \omega^{\cdot^{\cdot^{\cdot}}}}}}$$

which we write as

$$x = x' \pm \omega^{y'} \overset{\pm}{-} \omega^{z'} \overset{\pm}{-} \omega^{t'} \overset{\pm}{-} \omega^{\cdots}$$

so as to keep it on one line (almost).

It is important to realise that this expression does not determine x, since in fact there will always be many numbers with the same continued exponential. We shall only discuss this briefly and informally.

For the moment, let E stand for the formal expression

$$a \pm \omega^b \overset{\pm}{-} \omega^c \overset{\pm}{-} \omega^{\cdots}.$$

The first number to be born with this as its continued exponential will be called E_0, or just E. At later times, there will be constructed other numbers with this expression both to the left and right of E_0. The first of these will be called E_{-1} and E_1 respectively, and then E_2 will denote the first which is to the right of E_1, $E_{\frac{1}{2}}$ the first between E_0 and E_1, and so on, defining E_x for every number x.

The following examples will show why E_x exists for all numbers x. Let ε denote the particular expression with $a = b = c = \ldots = 0$ and all signs $+$, or more simply,

$$\varepsilon = \omega^{\omega^{\omega^{\cdot^{\cdot^{\cdot}}}}}.$$

Then ε_0 (or simply ε) denotes the first ordinal ε-number greater than ω, namely the number

$$\{\omega, \omega^{\omega}, \omega^{\omega^{\omega}}, \ldots \mid \},$$

and ε_1 denotes the next ε-number

$$\{\varepsilon + 1, \omega^{\varepsilon+1}, \ldots \mid \},$$

and so on. [This is the usual notation for the ordinal ε-numbers.]

What is ε_{-1}? This must be to the left of ε, and (being a leader) therefore to the left of $\varepsilon - 1$, thence of $\omega^{\varepsilon-1}, \omega^{\omega^{\varepsilon-1}}$, and so on. But considering the number

$$\delta = \{\text{ordinals} < \varepsilon \mid \varepsilon - 1, \omega^{\varepsilon-1}, \omega^{\omega^{\varepsilon-1}}, \ldots\}$$

we find it quite easy to prove that $\delta = \omega^{\delta}$, so that δ has indeed the continued exponential expression $\omega^{\omega^{\omega^{\cdot^{\cdot^{\cdot}}}}}$. Again, since in fact δ is the first number constructed left of ε with this expression, we have $\delta = \varepsilon_{-1} = [\omega^{\omega^{\omega^{\cdot^{\cdot^{\cdot}}}}}]_{-1}$. It seems reasonable to think of δ as the simplest ε-number which is *not* an ordinal number.

In a similar way, we see that the number $\varepsilon_{-\frac{1}{2}}$ is defined by

$$\varepsilon_{-\frac{1}{2}} = \{\delta + 1, \omega^{\delta+1}, \omega^{\omega^{\delta+1}}, \ldots \mid \varepsilon - 1, \omega^{\varepsilon-1}, \omega^{\omega^{\varepsilon-1}}, \ldots\}.$$

It is easy to show that these generalised ε-numbers are precisely the solutions of the equation $x = \omega^x$.

In a similar fashion we can show that the equation $x = \omega^{-x}$ has a *unique* solution

$$[\omega^{-\omega^{-\omega^{\cdot^{\cdot^{\cdot}}}}}]_0$$

and that more generally if we write

$$x = [\omega^{-\omega^{-\omega^{\cdot^{\cdot^{\cdot}}}}}]_z, \; y = [\omega^{-\omega^{-\omega^{-\cdot^{\cdot^{\cdot}}}}}]_{-z},$$

then we have $x = \omega^{-y}$, $y = \omega^{-x}$.

Of course these notations do not enable us to express every number in terms of earlier constructed numbers, since there will even be some numbers x associated with any given E which satisfy $x = E_x$. The first of these we should naturally call $E_{E_{E_{\cdot_{\cdot_{\cdot}}}}}$, the next born to the left of this being

$\left[E_{E_{E_{E_{\cdot_{\cdot_{\cdot}}}}}}\right]_{-1}$, and so on. But even before one makes the obvious extension

of our notation, there are other numbers to be considered, such as $E_{F_{G_{\cdot_{\cdot_{\cdot}}}}}$

where $E, F, G, \ldots$ are possibly different continued exponential expressions.

However, we now have a notation-system which is rich enough for all practical purposes, and is perhaps comparable with the usual system of notation for ordinal numbers.

SIGN-EXPANSIONS AND NORMAL FORMS

Consider a number $\Sigma\omega^y . r_y$ in normal form. How do we find the sign-expansion of this number given those of the y and r_y? We first cope with the effect of the condition that the terms are to be summed in descending order of y. We shall call the sign Y_δ in the sign-expansion $[Y_0, \ldots, Y_\delta, \ldots]$ of y *irrelevant* if the number with sign-expansion $[Y_0, \ldots, Y_\varepsilon, \ldots]_{\varepsilon<\delta}$ is greater than or equal to some $x > y$ with $r_x \neq 0$. Then the *relevant* sign-expansion of y is that obtained by omitting all the irrelevant signs from its ordinary sign-expansion.

Now suppose our number is written as $\omega^x . r + \omega^y . s + \omega^z . t + \ldots$ so as to display only the non-zero terms in its normal form. Then it turns out that its sign-expansion is obtained by juxtaposing those of $x, r, y, s, z, t, \ldots$ with each term repeated a power of ω times, except that the signs of $r, s, t, \ldots$ affect the entire expansions of $x, y, z, \ldots$, and irrelevant signs are omitted.

To be precise, suppose that $x, y, z, \ldots$ have relevant sign-expansions $[X_\delta]_{\delta<\alpha}$, $[Y_\delta]_{\delta<\beta}$, $[Z_\delta]_{\delta<\gamma}$, $\ldots$, and that $r, s, t, \ldots$ have (ordinary) sign-expansions $[R_0, R_1, \ldots]$, $[S_0, S_1, \ldots]$, $[T_0, T_1, \ldots]$, $\ldots$. Then the sign-expansion of $\omega^x . r + \omega^y . s + \omega^z . t + \ldots$ is

$$[(X_0R_0)^{\omega^{e_0+1}}, \ldots, (X_\delta R_0)^{\omega^{e_\delta+1}}, \ldots, R_1^{\omega^{e_\alpha}}, R_2^{\omega^{e_\alpha}}, \ldots,$$
$$(Y_0S_0)^{\omega^{f_0+1}}, \ldots, (Y_\delta S_0)^{\omega^{f_\delta+1}}, \ldots, S_1^{\omega^{f_\beta}}, S_2^{\omega^{f_\beta}}, \ldots,$$
$$(Z_0T_0)^{\omega^{g_0+1}}, \ldots, (Z_\delta T_0)^{\omega^{g_\delta+1}}, \ldots, T_1^{\omega^{g_\gamma}}, T_2^{\omega^{g_\gamma}}, \ldots, \ldots],$$

where for each $\delta < \alpha$, e_δ denotes the (ordinal) number of $+$ signs among the numbers X_ε ($\varepsilon < \delta$), and the numbers $f_\delta, g_\delta, \ldots$ are defined similarly for the numbers $y, z, \ldots$.

(The simplest proof is obtained by considering the sequence of successive approximations to $\omega^x . r + \omega^y . s + \omega^z . t + \ldots$ in their normal forms. Any such approximation is either a partial sum of the given normal form, or else differs from such a sum only in its final term.)

GAPS IN THE NUMBER LINE

Treading perhaps on rather thin ice, we now consider Dedekind sections (L, R) of **No** itself. Of course such a section, which we call a *gap*, consists of two disjoint Classes L, R whose union is **No**, with every member of R exceeding every member of L. If Ξ is this gap, and x a number, $x + \Xi$ is the gap $(x + L, x + R)$, $-\Xi$ the gap $(-R, -L)$, and ω^{Ξ} the gap (L', R') for which L' contains all numbers $\omega^{l}.s$ and R' all numbers $\omega^{r}.s$, where $l \in L$, $r \in R$, and s is any positive real.

Our theory of normal forms is easily extended to cater for gaps. In fact any gap has one of the two forms

$$\sum_{\beta \in \mathbf{On}} \omega^{x_\beta}.r_\beta \tag{1}$$

$$\sum_{\beta<\alpha} \omega^{x_\beta}.r_\beta \pm \omega^{\Xi'} \tag{2}$$

where in each case the sequence (x_β) is decreasing and each r_β is a non-zero real number, and in the second case Ξ' is a gap (L', R') for which R' contains all the x_β $(\beta < \alpha)$. In the first case the number $\Sigma\omega^{x_\beta}.r_\beta + \omega^{x_\alpha}.r'_\alpha$ $(r'_\alpha \neq r_\alpha)$ is in L or R according as $r'_\alpha < r_\alpha$ or $r'_\alpha > r_\alpha$.

The gaps definable as upper or lower bounds of sets are particularly important in the theory of games. It follows from the preceding remarks that any such gap has the form (2), where Ξ' is another gap of the same kind. Consequently we can continue, defining a sequence of numbers x_n and gaps Ξ_n so that $\Xi_n = x_n \pm \omega^{\Xi_{n+1}}$ and the gap $\Xi = \Xi_0$ has a continued exponential expression

$$\Xi = x_0 \pm \omega^{x_1 \pm \omega^{x_2 \pm \omega^{\cdots}}}.$$

The gap is not determined by this expression however—for instance $(\mathbf{No}, \varnothing)$ and the upper bound of all ordinals less than ε_0 both have the continued exponential $\omega^{\omega^{\omega^{\cdots}}}$.

Some gaps are important enough to deserve names. We use

"**On**" for the gap $(\mathbf{No}, \varnothing)$ at the end of the number line,

"$\frac{1}{\mathbf{On}}$" for the gap between 0 and all positive numbers,

"∞" for the gap between reals and positive infinite numbers, and

"$\frac{1}{\infty}$" for that between infinitesimals and the positive reals.

(This notation has been considerably extended in *Winning Ways*.)

For these gaps, we have

$$\mathbf{On} = \omega^{\mathbf{On}},\ \frac{1}{\mathbf{On}} = \omega^{-\mathbf{On}},\ \infty = \omega^{1/\mathbf{On}},\ \frac{1}{\infty} = \omega^{-1/\mathbf{On}},$$

from which we can deduce their continued exponentials.

As an example of a gap of the first kind, we give the normal form

$$\Xi* = 1 + \omega^{-1} + \omega^{-2} + \ldots + \omega^{-\omega} + \ldots + \omega^{-\alpha} + \ldots,$$

summed over *all* ordinals α, and as an example of a gap of the second kind that is not the upper or lower bound or a set we give $\omega^{\Xi*}$. There are also gaps of this sort with infinite continued exponentials, for example $\varepsilon_{\Xi*}$, in an obvious notation.

Just as we speak of an *infinity* of objects when the collection of them is not finite, it seems natural to speak of a *University* of objects when the Collection is a Proper Class. But the collection of all gaps is not even a Proper Class, being an illegal object in most set theories. Informally we may call it an IMPROPER CLASS, and speak of there being an *IMPROPRIETY* of gaps! There are very many gaps indeed. But we committed no impropriety in our discussion of them, which could all be formalised in such a way that at no point did the argument refer to more than one gap at a time.

Martin Kruskal has given a definition of $\exp(x)$ for all surreal numbers x, in which the options have the form 0 or $\exp(x') \cdot E_n(x - x')$, where $E_n(t) - 1 + t + \ldots + t^n/n!$ and there are some obvious restrictions on n and x'. It is easy to see that this function satisfies $\exp(x + y) = \exp(x) \cdot \exp(y)$ and is the inverse of the logarithm function that had been defined earlier using Simon Norton's integral of $1/x$, for which see the Epilogue.

Using these functions, we can define x^y in the usual way as $\exp(y \log(x))$. This *analytic power* has all the right properties, but the reader is warned that the ω-map of this chapter is not a particular case of it; for example, the number $\{1, \omega, \omega^2, \omega^3, \ldots \mid\}$ is $\exp(\omega) = e^\omega$ rather than the analytic ω^ω.

CHAPTER 4

Algebra and Analysis of Numbers

Now as to what pertains to these Surd numbers (which, as it were by way of reproach and calumny, having no merit of their own, are also styled Irrational, Irregular, and Inexplicable) they are by many denied to be numbers properly speaking, and are wont to be banished from Arithmetic to another Science (which yet is no science) viz., algebra.

Isaac Barrow, "Mathematical Lectures", 1734

In this chapter we show how the new numbers we have constructed are related to the real and complex numbers that are more familiar to the mathematician.

INFINITE SUMS

Suppose that to each number y we assign a real number r_y, with the restriction that r_y must vanish whenever y does not belong to a certain descending sequence $(y_\beta : \beta < \alpha)$. Then we define the formal sum $\sum_{y \in \mathrm{No}} \omega^y . r_y$ to be the value of the sum $\sum_{\beta < \alpha} \omega^{y_\beta} . r_\beta$ as defined in the previous chapter.

THEOREM 22.

$$\sum_{y \in \mathrm{No}} \omega^y . r_y + \sum_{y \in \mathrm{No}} \omega^y . s_y = \sum_{y \in \mathrm{No}} \omega^y . (r_y + s_y).$$

Proof. This follows easily from the formula

$$\sum_{y \in \mathrm{No}} \omega^y . r_y = \left\{ \sum_{y > z} \omega^y . r_y + \omega^z . r_z^L \,\middle|\, \sum_{y > z} \omega^y . r_y + \omega^z . r_z^R \right\}_{z \in \mathrm{No}}$$

The apparent appearance of a proper Class of numbers within the brackets is an illusion, since there is no r_z^L or r_z^R for z outside a certain set.

Summary. The formal sums we encounter when dealing with normal forms have properties compatible with finite sums.

This result allows us to define various more general infinite sums. In general we shall write the normal form of a number x in the form $\sum_y \omega^y . r_y$, it being understood that the sum is over all $y \in \mathbf{No}$, and that the numbers r_y satisfy the required conditions. If now we have a set or sequence of numbers $x_n = \sum_y \omega^y . r_{n,y}$, then we say that the sum $\sum_n x_n$ is *convergent* to x (in some sense) if and only if all the real number sums $\sum_n r_{n,y}$ are convergent (in the same sense) to sums r_y, say, and x is the number $\Sigma\, \omega^y . r_y$, and *furthermore* all the $r_{n,y}$ vanish for all y not in some descending sequence $(y_\beta : \beta < \alpha)$. This last restriction is quite essential to prevent certain absurdities—without it we should have

$$(1 - \omega) + (\omega - \omega^2) + (\omega^2 - \omega^3) + \ldots = 1,$$

in which an infinite sequence of negative numbers has positive sum. We call a number *infinitesimal* if it lies between $-r$ and r for every positive real number r.

THEOREM 23. *A power series with real coefficients is always absolutely convergent for all infinitesimal values of the variable.*

Proof. This requires only the (rather subtle) theorem that if $\{y_\beta\}$ is a reversely well-ordered subset of negative numbers in a totally ordered group, then so is the set of all finite sums of the y_β. We quote this theorem. [A direct proof is not hard, but the theorem really belongs to a large Class(!) of combinatorial theorems about well-ordered sets which do not really concern us here. The particular result we want is proved, for essentially the same application, in Cohn's treatment of the Malcev–Neumann construction.]

THEOREM 24. *Every positive number x has a unique positive n-th root, for each positive integer n.*

Proof. By considering the normal form of x, we see that we can write

$$x = \omega^y . r . (1 + \delta)$$

where δ is some infinitesimal number. Then the number

$$\omega^{y/n} . r^{1/n} . \left[1 + \frac{1}{n} \cdot \delta + \frac{1}{n} \cdot \left(\frac{1}{n} - 1\right) \frac{\delta^2}{2} + \ldots\right]$$

is an nth root of x. The uniqueness is obvious.

ROOTS OF ODD-DEGREE POLYNOMIALS

Let $f(x) = x^n + Ax^{n-1} + Bx^{n-2} + \ldots + K$ be some polynomial of odd degree. We intend to show that f has a root in the Field of all numbers. By translating x by a suitable number, we can suppose that $A = 0$. Then unless the polynomial is the rather trivial polynomial x^n (which has the root 0), we can replace x by tx for some number t so as to ensure that

$$\max(|B|, |C|, \ldots, |K|) = 1.$$

Then $f(x)$ takes the form

$$x^n + (b + \beta)x^{n-2} + (c + \gamma)x^{n-3} + \ldots + (k + \kappa),$$

in which $b, c, \ldots, k$ are real numbers not all zero, and $\beta, \gamma, \ldots,$ denote not ordinals but infinitesimal numbers.

For a first approximation, put $\beta = \gamma = \ldots = 0$. Then the resulting polynomial has a factorisation

$$(x - p)^{n_1} . (x - q)^{n_2} \ldots (x - r)^{n_j}$$

in which $p, q, \ldots, r$ are distinct complex numbers. Moreover, since the sum of the roots is zero, we must have $j \geqslant 2$.

We can now group the complex numbers in conjugate pairs to obtain a factorisation

$$f_1(x) . f_2(x) \ldots f_J(x)$$

in which the f_t are polynomials with *real* coefficients, and no two of the f_t have a common root. Moreover, we still have $J \geqslant 2$ since the degree of f was odd.

Now put back the numbers $\beta, \gamma, \ldots, \kappa$, but regard them for the moment as small complex numbers. Then the perturbed polynomial has a corresponding factorisation whose coefficients are analytic functions of $\beta, \gamma, \ldots, \kappa$, which can therefore be expressed as power-series in these variables, convergent for sufficiently small values of them. (The assumption that the f_t have roots distinct from each other is needed to prevent these analytic functions from having branch-points at the origin.) These power-series will certainly converge for *infinitesimal* values of $\beta, \gamma, \ldots, \kappa$, and so we obtain a non-trivial factorisation of f whose coefficients are numbers. So we have

THEOREM 25. *Every odd-degree polynomial with coefficients in* **No** *has a root in* **No**.

Proof. This follows from the above argument by induction, since at least one of the factor polynomials will still have odd degree.

Now Gauss' third proof of the so-called fundamental theorem of algebra shows essentially that if we have any field in which for each x either x or $-x$ has a square root, and every odd degree polynomial has a root, then we obtain an algebraically closed field by adjoining a square root i of -1. (Artin has made this the basis of his elegant theory of *real-closed fields.*)

So we have:

THEOREM 26. *The Ring* **No**$[i]$ *of all "numbers" of the form* $x + iy$ $(x, y \in \mathbf{No})$, $i^2 = -1$, *is an algebraically closed Field.*

If we do not wish to adjoin i, we may make effectively the same assertion by saying that **No** is itself a real-closed Field.

Using the axiom of choice, it is quite easy to see that **No**$[i]$ is as an abstract Field the algebraic closure of the Field obtained from $\mathbb{Q}$ by adjoining a "University" of independent transcendentals (that is, one for each member of the Universe). A theorem of Artin's now enables us to deduce a characterisation of **No** as an abstract Field. Summing up:

THEOREM 27. **No**$[i]$ *is characterised up to Field isomorphism by the fact that it is the algebraic closure of the rationals extended by a University of transcendentals.*

And now Artin's theorem asserts that any field whose algebraic closure is only a finite extension is either algebraically closed or real-closed, in the latter case under an order which is unique up to field isomorphism.

Of course **No** has much more structure than this, so that Theorem 27 is in no sense a substitute for the definition of **No**. For when we consider it together with its collection structure $\{L \mid R\}$, **No** has plainly only the identity automorphism. We now give an alternative characterisation of **No** as a totally ordered Field.

We say that a Field F (necessarily a proper Class) is *universally embedding* if each (set) subfield f of F which as an abstract totally ordered field can be extended to a field g is already contained in a subfield $\bar{g}$ of F isomorphic to g, the isomorphism restricting to the identity on f.

THEOREM 28. **No** *is a universally embedding totally ordered Field.*

Proof. (The proof uses the axiom of choice.) We need only tackle the inductive step, which is when g is obtained from f by real-closure after the adjunction of a single new element x. We consider all polynomials in x with coefficients in f. Then using the real-closure of **No** and the fact that every section of f contains points of **No** we can produce an $\bar{x}$ in **No** for which the corresponding polynomials have the same signs as at x. We take $\bar{g}$ as f extended by $\bar{x}$, with the isomorphism taking x to $\bar{x}$, and then real-close so as to preserve the order.

Finally, we see that this property again defines **No** as an abstract Field.

THEOREM 29. *Any universally embedding totally ordered Field is isomorphic to* **No**.

Proof. The proof uses the fact that all proper Classes have the same "Cardinal", which follows from the axioms of choice and foundation (and which was taken as an axiomatic *definition* of proper Class by von Neumann!) Using this, we can well-order the elements of **No** and of F in order-type **On**. We first identify their rational subfields, and then "patch up" an isomorphism by alternately finding images of elements of the two Fields inside each other, real-closing after each adjunction, and always taking the first element (in the appropriate well-ordering) not yet dealt with.

Summary. As an abstract Field, **No** is the unique universally embedding totally ordered Field.

We repeat that **No** has plenty of additional structure which would not emerge from this "definition".

FURTHER REMARKS ABOUT ANALYSIS IN No

The theory of infinite sums enables us to do quite a lot of classical analysis in **No** (or often more easily in **No**$[i]$). Thus various analytic functions can be defined on large parts of **No** by power-series whenever these are convergent. The exponential series converges for example whenever $|x| <$ some finite integer n, and defines a perfectly respectable number-function with the expected properties inside this region. Similarly we can define sines and cosines, etc., in the same region. Logarithms can be defined in the same region (except at infinitesimal x) by means of the power-series for $\log(1 + t)$ and the formula $\log(nx) = \log n + \log x$, where n is an ordinary positive integer.

The exponential and logarithmic functions obtained by this general method agree (in this region) with the everywhere-defined ones mentioned at the end of Chapter 3. But the problem of defining other classical functions outside this "bounded" region has still not been solved, although some progress has been made by M. D. Kruskal.

It is interesting to note that our definitions of infinite sums have in a certain sense to be "global", rather than as limits of partial sums, because limits don't seem to work. For instance, the limit of the sequence $0, \frac{1}{2}, \frac{2}{3}, \frac{3}{4}, \ldots$ (ω terms) is not 1, at least in the ordinary sense, because there are plenty of numbers in between. A simpler, but sometimes less convincing, example of the same phenomenon is given by the sequence

$$0, 1, 2, 3, \ldots$$

of all finite ordinals, which one would expect to tend to ω, but which obviously can't, since there is a whole Host of numbers greater than every finite integer but less than ω. For the author's amusement, we recall some of the simplest of them:

$$\omega - 1,\ \omega/2,\ \sqrt{\omega},\ \omega^{1/\omega},\ [\omega^{\omega^{-\omega^{-\omega}\cdots}}]_x \text{ (for all } x!\text{)},$$
$$\omega^{\omega^{-\omega}},\ \omega^{\omega^{-\varepsilon}},\ \ldots.$$

NON-STANDARD ANALYSIS

We can of course use the Field of all numbers, or rather various small subfields of it, as a vehicle for the techniques of non-standard analysis developed by Abraham Robinson. Thus for instance for any reasonable function f, we can define the derivative of f at the real number x to be the closest real number to the quotient

$$\frac{f[x + (1/\omega)] - f(x)}{1/\omega}.$$

The reason is that *any* totally ordered real-closed field is a model for the elementary statements about the real numbers. But for precisely this reason, there is little point in using subfields of **No** when so many more visible fields will do. So we can say that in fact the Field **No** is really irrelevant to non-standard analysis.

[The reader might be tempted to suppose that the subRing of *omnific integers* described in the next chapter was in a similar way a non-standard Model for the ordinary integers. But of course this is not so, since for instance $x^2 = 2y^2$ has many non-zero omnific integer solutions. In fact deep logical theorems tell us that we could not hope to find a non-standard model for $\mathbb{Z}$ in so simple a way.]

CHAPTER 5

Number Theory in the Land of Oz

"We're off to see the Wizard,
The Wonderful Wizard of Oz!"
After title of book by L. Frank Baum.

In this chapter we discuss the notion of *integer* which is appropriate to our big Field **No**.

Definition. [Norton]. x is an *omnific integer* iff $x = \{x - 1 \mid x + 1\}$. We shall use **Oz** for the Class of omnific integers. In this chapter the unqualified word integer will usually mean omnific integer.

THEOREM 30. (i) 0 *is an integer,*
(ii) *if* x *is an integer, so is* $-x$,
(iii) *if* x *and* y *are integers, so are* $x + y$ *and* xy.

Proof. These have 1-line proofs.

THEOREM 31. *The number* $x = \Sigma\, \omega^y . r_y$ *is in* **Oz** *if and only if* $r_y = 0$ *for* $y < 0$, *while* $r_0 \in \mathbb{Z}$. *Every ordinal number is an integer.*

Proof. Define $\bar{r}_y = r_y$ for $y > 0$, 0 for $y < 0$, and $\{r_0 - 1 \mid r_0 + 1\}$ for $y = 0$. Then the number $\bar{x} = \Sigma\, \omega^y . \bar{r}_y$ certainly lies between $x - 1$ and $x + 1$, and is simpler than or equal to x. So x is an integer if and only if $\bar{x} = x$. The second sentence now follows.

THEOREM 32. *Every number* x *is the quotient of two omnific integers.*

Proof. Taking x as above, with $r_y = 0$ for $y \leqslant -\alpha$ (α some ordinal, say), we observe that ω^α and $x\omega^\alpha$ are both integers by Theorem 31.

So for example the number π is the quotient of the two integers $\omega\pi$ and ω.

Summary. **Oz** includes **On** and is a subRing of **No**, with **No** as its Field of quotients. Every number is distant at most 1 from some omnific integer.

Definition. An integer is *divisible* if and only if it is divisible by every finite non-zero integer.

THEOREM 33. *Each integer is uniquely the sum of a divisible and a finite integer.*

Proof. If x is the integer $\Sigma\, \omega^y . r_y$, then r_0 is finite and $x - r_0$ divisible. If r is any finite integer with $x - r$ divisible, then $r - r_0$ is divisible and so $r = r_0$.

THEOREM 34. *If a and b are integers with b positive, there are unique integers q and r with* $a = bq + r, 0 \leqslant r < b$.

Proof. Let $x = a/b$, and $\bar{x}$ the integer $\{x - 1 \mid x + 1\}$, so that

$$a - b < b\bar{x} < a + b.$$

Then if $a - b < b\bar{x} \leqslant a$ we can take $q = \bar{x}$, and otherwise $q = \bar{x} - 1$. Inequalities imply the uniqueness.

When restricted to ordinal numbers Theorem 33 becomes essentially the particular case $b = \omega$ of Theorem 34. But for general numbers they are quite distinct theorems.

Since there is no descending chain condition for omnific integers, Theorem 34 does *not* show that these integers have unique factorisation. In fact, for example, ω has infinitely many distinct factorisations

$$\omega = 2 . \frac{\omega}{2} = 3 . \frac{\omega}{3} = \ldots = (\sqrt{\omega})^2, \textit{ etc.}$$

The same thing can happen for indivisible integers, for example

$$\omega + 1 = (\omega^{\frac{1}{3}} + 1)(\omega^{\frac{2}{3}} - \omega^{\frac{1}{3}} + 1) = (\omega^{\frac{1}{5}} + 1)(\omega^{\frac{4}{5}} - \ldots + 1) = \ldots .$$

But certain other infinite integers appear to be prime, for instance

$$\omega + \omega^{\frac{1}{2}} + \omega^{\frac{1}{3}} + \ldots + 1.$$

Conjecture. Omnific integers have *the refinement property*—if $ab = cd$ for omnific integers, then there are further integers e, f, g, h with $a = ef$, $b = gh$, $c = eg$, $d = fh$.

WARING'S PROBLEM

At first sight one is inclined to think that perhaps every divisible integer is, like ω, a perfect nth power for every finite integer n. But the divisible integer $\omega^2 + \omega$ is not even a square, for it lies between the squares of the adjacent integers ω and $\omega + 1$. But $\omega^2 + \omega$ is the sum of *two* squares, namely those of ω and $\sqrt{\omega}$. However, $\omega^2 - 1$ is not the sum of *any* number of squares, for then their finite parts would be finite squares summing to -1. So Waring's problem fails for squares.

If we allow cubes of negative integers, we can prove, however, that every integer is the sum of at most five cubes, by imitating the standard proof for finite integers.

For we have the identity

$$(x - 1)^3 + (-x)^3 + (-x)^3 + (x + 1)^3 = 6x,$$

and using Theorem 34 we can write any integer in the form $6x - n^3$, where x is integral and $n = 0, 1, 2, 3, 4,$ or 5, since the cubes of these numbers exhaust the residue classes modulo 6. Just as in the finite case we see that 4 cubes are sometimes needed (consider numbers congruent to ± 4 modulo 9), but again just as in the finite case we do not know whether 4 will always suffice.

CONTINUED FRACTIONS AND PELLIAN EQUATIONS

If x is a positive number there is an integer $[x]$ so that $[x] \leqslant x < [x] + 1$. Of course $[x]$ is called the *integer part* of x. Let a be the integer part of x, and if $x \neq a$ write $x = a + (1/y)$. Then if y is distinct from its integer part b, write $y = b + (1/z)$, and so on. The process may terminate at a finite stage if one of the remainders $y, z, \ldots$ is zero, but otherwise we obtain an infinite *continued fraction* (we use the standard abbreviated notation)

$$a + \frac{1}{b +} \frac{1}{c +} \frac{1}{d +} \cdots$$

corresponding to x.

[Those x for which the continued fraction does terminate at some finite stage are naturally called *fractional numbers*, since they are perhaps the closest analogue in **No** of the ordinary rational numbers. If x is fractional, so are $x + 1$, $-x$, and $1/x$ (if $x \neq 0$), but neither the sum nor the product of two fractional numbers need be fractional—consider

$$\frac{1}{\omega}, \frac{1}{\pi\omega^2 + \omega}, \quad \text{and} \quad \omega\sqrt{2}, \ \frac{1}{\omega}.$$

The equation $x^2 - Ny^2 = \pm 1$, where N is a given integer and x and y are to be found as omnific integers, is readily discussed in terms of continued fractions. Almost exactly as in the finite case, we can show that x/y must be one of the convergents to the continued fraction for $\sqrt{N}$. (It is essential to note that there cannot exist any solution in which x and y are too large compared with N.) Thus the equation has at most $\aleph_0$ solutions.

Sometimes the form of the solutions is quite surprising. We consider for example the case $N = \omega + 3$.

Here we find successively

$$\sqrt{(\omega + 3)} = \sqrt{\omega} + \frac{1}{u}, \text{ say}$$

$$u = \tfrac{1}{3}(\sqrt{(\omega + 3)} + \sqrt{\omega}) = \tfrac{2}{3}\sqrt{\omega} + \frac{1}{v}, \text{ say}$$

$$v = \sqrt{(\omega + 3)} + \sqrt{\omega} = 2\sqrt{\omega} + \frac{1}{u}$$

so that $\sqrt{(\omega + 3)}$ yields the periodic continued fraction

$$\sqrt{\omega} + \frac{1}{\tfrac{2}{3}\sqrt{\omega} +}\,\frac{1}{2\sqrt{\omega} +}\,\frac{1}{\tfrac{2}{3}\sqrt{\omega} + \dots}$$

whose first few convergents are

$$\frac{\sqrt{\omega}}{1}, \quad \frac{\tfrac{2}{3}\omega + 1}{\tfrac{2}{3}\sqrt{\omega}}, \quad \frac{\tfrac{4}{3}\omega\sqrt{\omega} + 3\sqrt{\omega}}{\tfrac{4}{3}\omega + 1}, \quad \frac{\tfrac{8}{9}\omega^2 + \tfrac{8}{3}\omega + 1}{\tfrac{8}{9}\omega\sqrt{\omega} + \tfrac{4}{3}\sqrt{\omega}}, \dots .$$

The alternate ones of these do indeed yield solutions of the equation, namely

$$(\tfrac{2}{3}\omega + 1)^2 - (\omega + 3)(\tfrac{2}{3}\sqrt{\omega})^2 = 1$$

$$(\tfrac{8}{9}\omega^2 + \tfrac{8}{3}\omega + 1)^2 - (\omega + 3)(\tfrac{8}{9}\omega\sqrt{\omega} + \tfrac{4}{3}\sqrt{\omega})^2 = 1$$

$$\dots$$

as can easily be checked.

Not all such equations behave so exactly like the finite case. Although the square roots of many simple integers yield periodic continued fractions, there are some that do not, for instance $\sqrt{(\omega^2 + 2e\omega)}$ (where e is the base of the natural logarithms) yields the same continued fraction as $\omega + e$, namely

$$\omega + 2 + \frac{1}{1 +}\,\frac{1}{2 +}\,\frac{1}{1 +}\,\frac{1}{1 +}\,\frac{1}{4 +} \dots .$$

Plainly no convergent of this leads to a solution of the corresponding Pellian equation, which is therefore insoluble. Other behaviours are possible. (Note in passing that the continued fraction of a number does not always determine that number. There does not seem to be any way of extending the definition so as to define partial quotients for the ωth stage and beyond.)

Almost every number-theoretical problem can be rephrased so as to yield a new problem in **Oz**, so we get a jackdaw's nest of problems of various kinds. But it seems in general that problems whose usual solution involves

Gauss's theory of congruences tend to produce rather trivial generalisations in **Oz**, while those whose normal treatment involves rational approximation or cunning algebraic identities produce more interesting problems.

Often we can get even more interesting problems by generalising ordinary problems so as to allow infinitely many variables. We mention only one before finishing this rather short chapter:

Is every positive omnific integer the sum of a number (possibly infinite) of positive perfect cubes (of omnific integers)?

The scarecrow will need to take some time to think before giving his answer.

CHAPTER 6

The Curious Field $\mathbf{On}_2$

"The way into my parlour is up a winding stair,
And I have many curious things to show when you are there."
Mary Howitt, "The Spider and the Fly"

The main idea of this Chapter is that we abolish the distinction between L and R (and so between $+$ and $-$), and explore the consequences of our genetic definitions of arithmetic operations in this more symmetrical context. What we get is in a sense the characteristic 2 analogue of the big Field **No**, which we might naturally call $\mathbf{No}_2$. But it turns out that this new Field is also the "simplest" way of turning the Class **On** of all ordinal numbers into a Field, and so for a moment we shall explore it from this viewpoint and adopt the name $\mathbf{On}_2$ (which has in any case a nicer sound).

How shall we find the *simplest* addition and multiplication which make **On** a Field? (The reader who is happier with integers than with general ordinals can restrict his attention to the non-negative integers $0, 1, 2, 3, \ldots\ldots$) We might do this as follows. We first fill in the addition-table, subject to the condition that before we fill in the entry for $\alpha + \beta$ we must have already filled in all entries $\alpha' + \beta$ and $\alpha + \beta'$ ($\alpha' < \alpha$, $\beta' < \beta$). Then the entry at $\alpha + \beta$ is to be the least possible number which is consistent with the result's being part of the addition-table of a Field. We then tackle the multiplication-table of a Field with the given addition. Again, the entries are to be the least possible ones subject to this requirement.

In this way we obtain the tables of Figs 2 and 3. We discuss the first few entries.

We have $0 + 0 = 0$, since 0 is the least conceivable value, and there certainly is a field with an element satisfying $x + x = x$, namely *any* field, with x as the zero element. But then this equation implies that 0 must be the zero element of our Field, and so we must have $0 + \alpha = \alpha + 0 = \alpha$ for all α.

What about $1 + 1$? The least conceivable answer is 0, for there exist fields of characteristic 2. So we must have $1 + 1 = 0$, and so $\alpha + \alpha = 0$ for all α.

The next entry is $1 + 2$. This must be distinct from 0, 1, and 2, and so can and must be taken as 3. We then have $1 + 3 = 1 + 1 + 2 = 2$, $2 + 3 = 2 + 1 + 2 = 1$, and we know all sums $\alpha + \beta$ with both α and β less than 4. We must have $4 + 0 = 4$, $4 + 1 = 5$, $4 + 2 = 6$, and $4 + 3 = 7$ since these numbers must all be distinct from 0, 1, 2, 3. Using these, we can fill in all sums $\alpha + \beta$ with α and γ less than 8, and then we must have $8 + 0 = 8$, $8 + 1 = 9, \ldots, 8 + 7 = 15$, yielding all sums of numbers less than 16, and so on. So the addition-table is, in part:

+	0	1	2	3	4	5	6	7	8	9	10	11	12	13	14	15
0	0	1	2	3	4	5	6	7	8	9	10	11	12	13	14	15
1	1	0	3	2	5	4	7	6	9	8	11	10	13	12	15	14
2	2	3	0	1	6	7	4	5	10	11	8	9	14	15	12	13
3	3	2	1	0	7	6	5	4	11	10	9	8	15	14	13	12
4	4	5	6	7	0	1	2	3	12	13	14	15	8	9	10	11
5	5	4	7	6	1	0	3	2	13	12	15	14	9	8	11	10
6	6	7	4	5	2	3	0	1	14	15	12	13	10	11	8	9
7	7	6	5	4	3	2	1	0	15	14	13	12	11	10	9	8
8	8	9	10	11	12	13	14	15	0	1	2	3	4	5	6	7
9	9	8	11	10	13	12	15	14	1	0	3	2	5	4	7	6
10	10	11	8	9	14	15	12	13	2	3	0	1	6	7	4	5
11	11	10	9	8	15	14	13	12	3	2	1	0	7	6	5	4
12	12	13	14	15	8	9	10	11	4	5	6	7	0	1	2	3
13	13	12	15	14	9	8	11	10	5	4	7	6	1	0	3	2
14	14	15	12	13	10	11	8	9	6	7	4	5	2	3	0	1
15	15	14	13	12	11	10	9	8	7	6	5	4	3	2	1	0

FIG. 2. Nim-addition.

Readers familiar with the theory of the game of Nim will recognise this operation as the addition used in that game, so we refer to it as *Nim-addition*. The following is an easy rule enabling us to perform Nim-additions:

(i) The Nim-sum of a number of distinct 2-powers is their ordinary sum. Thus $8 + 4 + 1$ is still 13.

(ii) The Nim-sum of two equal numbers is 0.

We use the term *2-power* to mean a power of 2 in the ordinary sense, such as $1, 2, 4, 8, 16, \ldots$. (These are not powers of 2 with the new multiplication.)

Using this Nim-addition is easy, for example

$$13 + 7 = (8 + 4 + 1) + (4 + 2 + 1) = 8 + 2 = 10,$$

since the 4's and 2's cancel. The rule is of course the same as the usual rule "write the numbers down in binary and then add without carrying", but we find that with that rule there are far too many opportunities to make mistakes while making the unnecessary translations.

We shall give a formal proof of the rule later.

With multiplication, we find that $0.\alpha$ *can* and so *must* be 0, so that 0 must be the zero of the Field. Then 1.1 can and so must be 1, so that 1 is the one, which results enable us to fill the first two rows and columns. We next observe that 2.2 cannot be 0, 1, or 2, but can be 3, since in the finite field of order 4, the elements other than 0 and 1 satisfy $x^2 = x + 1$. Similar but more complicated considerations give Fig. 3 as the first part of the multiplication-table:

.	0	1	2	3	4	5	6	7	8	9	10	11	12	13	14	15
0	0	0	0	0	0	0	0	0	0	0	0	0	0	0	0	0
1	0	1	2	3	4	5	6	7	8	9	10	11	12	13	14	15
2	0	2	3	1	8	10	11	9	12	14	15	13	4	6	7	5
3	0	3	1	2	12	15	13	14	4	7	5	6	8	11	9	10
4	0	4	8	12	6	2	14	10	11	15	3	7	13	9	5	1
5	0	5	10	15	2	7	8	13	3	6	9	12	1	4	11	14
6	0	6	11	13	14	8	5	3	7	1	12	10	9	15	2	4
7	0	7	9	14	10	13	3	4	15	8	6	1	5	2	12	11
8	0	8	12	4	11	3	7	15	13	5	1	9	6	14	10	2
9	0	9	14	7	15	6	1	8	5	12	11	2	10	3	4	13
10	0	10	15	5	3	9	12	6	1	11	14	4	2	8	13	7
11	0	11	13	6	7	12	10	1	9	2	4	15	14	5	3	8
12	0	12	4	8	13	1	9	5	6	10	2	14	11	7	15	3
13	0	13	6	11	9	4	15	2	14	3	8	5	7	10	1	12
14	0	14	7	9	5	11	2	12	10	4	13	3	15	1	8	6
15	0	15	5	10	1	14	4	11	2	13	7	8	3	12	6	9

FIG. 3. Nim-multiplication.

The entries in the printed part of the table can all be found from those we have already established and the further entries $4.2 = 8$, $4.4 = 6$, so we shall rapidly justify these. As for 4.2, this cannot be 0, 1, 2, or 3, since we already know that these numbers form a subfield not containing 4. Similarly 4.2 cannot be one of 4, 5, 6, or 7, since this would make 4.3 one of 0, 1, 2, or 3. Since all later numbers are essentially equivalent, 4.2 can and so must be taken as 8. Now 4.4 cannot be one of 0, 1, 2, 3 since these numbers are already squares in $\{0, 1, 2, 3\}$, and a number cannot have more than one square root in a field of characteristic 2. The equation $4.4 = 4$ would imply $4 = 1$, and $4.4 = 5$ would imply $4^2 + 4 = 1$, whereas the quadratic equation $x^2 + x = 1$ has already its full complement of roots (2 and 3) in the field $\{0, 1, 2, 3\}$. So

4.4 is at least 6, and since in fact the displayed multiplication-table does actually define a field of order 16, 4.4 can and must be 6.

[We could be rather less bold and simply assert that the equation $x^2 = x+2$ is irreducible over $\{0, 1, 2, 3\}$ and so we could adjoin a solution of it to obtain a larger field. This solution can, and so may, be called 4.]

It can be shown that for finite numbers the Nim-multiplication table follows from the following rules, analogous to those for Nim-addition. We shall use the term *Fermat 2-power* to denote one of the numbers 2, 4, 16, 256, 65536, ..., that is to say, the numbers 2^{2^n} in the ordinary sense.

(i) The Nim-product of a number of distinct Fermat 2-powers is their ordinary product. Thus 16.4.2 is still 128.

(ii) The square of a Fermat 2-power is its *sesquimultiple*.

The *sesquimultiple* of a number is the number obtained by multiplying it by $1\frac{1}{2}$ in the ordinary sense. So $2^2 = 3$, $4^2 = 6$, $16^2 = 24, \ldots$.

To work out the products of other numbers we use the associative and distributive laws. For example

$$5.9 = (4 + 1)(4.2 + 1) = 4^2.2 + 4.2 + 4 + 1 = 6.2 + 8 + 4 + 1$$
$$= (4 + 2).2 + 13 = 4.2 + 2^2 + 13 = 8 + 3 + 13 = 6.$$

Our two rules for addition and multiplication imply and are implied by the following rules, which are remarkably similar to each other:

(a) If x is a 2-power, and $y < x$, then $x + y$ has its normal value, but $x + x = 0$.

(b) If x is a Fermat 2-power, and $y < x$, then xy has its normal value, but $x.x$ is the ordinary value of $3x/2$.

The rule we have given for addition generalises to infinite ordinal numbers in a fairly obvious way, but that for multiplication does not, and we obtain many remarkable results, for instance the theorem that the least infinite ordinal ω is a cube root of 2!

THE INDUCTIVE DEFINITIONS

The definition of the operations in the above discussion is not very easy to work with, for we must prove a theorem every time we want to fill in an entry. In any case, it is not at all obvious that the definition is in any sense consistent, in the sense that it really does define a Field. It is remarkable that precisely the same effect is achieved by making just two 1-line definitions:

$\alpha + \beta$ *is the least ordinal distinct from all numbers* $\alpha' + \beta$, $\alpha + \beta'$

[$-\alpha$ *is the least ordinal distinct from all numbers* $-\alpha'$]

$\alpha\beta$ *is the least ordinal distinct from all numbers* $\alpha'\beta + \alpha\beta' - \alpha'\beta'$.

In each case, α' and β' represent arbitrary ordinals smaller than α and β respectively. We say *two* 1-line definitions because in fact $-\alpha = \alpha$ for all α, so that we could replace the $-$ sign by a $+$ sign in the product definition and eliminate the middle line. But we prefer to use $-$ signs where they seem more natural than $+$ signs.

In the formal development which follows, we shall use only these definitions. It will turn out that they do in fact make the Class **On** of ordinal numbers into a Field $\mathbf{On}_2$ with many curious properties. We hope the analogy with the definitions of the operations in **No** will not have escaped the reader. [*Least* really means *simplest.*]

We shall write mex(S) (minimal excluded number) for the least ordinal not in the set S, and refer to the members of S as *excludents*. If $\alpha = \text{mex}(S)$, we shall often use $\alpha*$ for a variable ranging over S—thus $\alpha*$ may take *all* values *less* than α and possibly *some* values greater than α, but not α itself. We continue to use α' for a variable which takes all values less than α.

PROPERTIES OF ADDITION

THEOREM. 36. *We have* $\alpha + \beta = \alpha + \gamma$ iff $\beta = \gamma$. *Also,*

$$\alpha + \beta = \text{mex}\{\alpha* + \beta, \alpha + \beta*\}.$$

Proof. If, say, $\beta > \gamma$, then $\alpha + \gamma$ is an excludent for $\alpha + \beta$. The second sentence follows, for certainly all numbers $\alpha' + \beta, \alpha + \beta'$ are excludents, and the other excludents are distinct from $\alpha + \beta$.

THEOREM. 37. *For all ordinals* α, β, γ *we have*

$$\alpha + 0 = \alpha, \quad \alpha + \beta = \beta + \alpha, \quad (\alpha + \beta) + \gamma = \alpha + (\beta + \gamma),$$
$$\alpha + \alpha = 0, \quad -\alpha = \alpha.$$

Proof. These have 1-line proofs:

$$\alpha + 0 = \text{mex}\{\alpha' + 0, \alpha + 0'\} = \text{mex}\{\alpha'\} = \alpha$$

$$\alpha + \beta = \text{mex}\{\alpha' + \beta, \alpha + \beta'\} = \text{mex}\{\beta + \alpha', \beta' + \alpha\} = \beta + \alpha$$

$$\begin{aligned}(\alpha + \beta) + \gamma &= \text{mex}\{(\alpha + \beta)* + \gamma, (\alpha + \beta) + \gamma'\}\\ &= \text{mex}\{(\alpha' + \beta) + \gamma, (\alpha + \beta') + \gamma, (\alpha + \beta) + \gamma'\}\\ &= \text{mex}\{\alpha' + (\beta + \gamma), \alpha + (\beta' + \gamma), \alpha + (\beta + \gamma')\}\\ &= \ldots = \alpha + (\beta + \gamma).\end{aligned}$$

$$\alpha + \alpha = \text{mex}\{\alpha' + \alpha, \alpha + \alpha'\} = \text{mex}\{0*\} = 0$$

$$-\alpha = \text{mex}\{-\alpha'\} = \text{mex}\{\alpha'\} = \alpha.$$

(Note the occasional occurrences of $*$.)

Summary. $\mathbf{On}_2$ forms an Abelian Group with 0 for zero and $-\alpha = \alpha$.

PROPERTIES OF MULTIPLICATION

THEOREM 38. *For all ordinals* α, β, γ *we have*

$$\alpha 0 = 0, \quad \alpha 1 = \alpha, \quad \alpha\beta = \beta\alpha, \quad (\alpha + \beta)\gamma = \alpha\gamma + \beta\gamma, \quad (\alpha\beta)\gamma = \alpha(\beta\gamma).$$

Proof. These also have 1-line proofs:

$$\alpha 0 = \text{mex}\{\ \} = 0$$

$$\alpha 1 = \text{mex}\{\alpha' 1 + \alpha 0 - \alpha' 0\} = \text{mex}\{\alpha'\} = \alpha$$

$$\alpha\beta = \text{mex}\{\alpha'\beta + \alpha\beta' - \alpha'\beta'\} = \text{mex}\{\beta'\alpha + \beta\alpha' - \beta'\alpha'\} = \beta\alpha$$

$$\begin{aligned}(\alpha + \beta)\gamma &= \text{mex}\{(\alpha + \beta)*\gamma + (\alpha + \beta)\gamma' - (\alpha + \beta)*\gamma'\} \\ &= \text{mex}\{(\alpha' + \beta)\gamma + (\alpha + \beta)\gamma' - (\alpha' + \beta)\gamma', \\ &\quad (\alpha + \beta')\gamma + (\alpha + \beta)\gamma' - (\alpha + \beta')\gamma'\} \\ &= \text{mex}\{(\alpha'\gamma + \alpha\gamma' - \alpha'\gamma') + \beta\gamma, \alpha\gamma + (\beta'\gamma + \beta\gamma' - \beta'\gamma')\} \\ &= \text{mex}\{(\alpha\gamma)* + \beta\gamma, \beta\gamma + (\beta\gamma)*\} = \alpha\gamma + \beta\gamma.\end{aligned}$$

$$\begin{aligned}(\alpha\beta)\gamma &= \text{mex}\{(\alpha\beta)*\gamma + (\alpha\beta)\gamma' - (\alpha\beta)*\gamma'\} \\ &= \text{mex}\{(\alpha'\beta + \alpha\beta' - \alpha'\beta')\gamma + (\alpha\beta)\gamma' - (\alpha'\beta + \alpha\beta' - \alpha'\beta')\gamma'\} \\ &= \text{mex}\{\alpha'\beta\gamma + \alpha\beta'\gamma + \alpha\beta\gamma' - \alpha'\beta'\gamma - \alpha'\beta\gamma' - \alpha\beta'\gamma' + \alpha'\beta'\gamma'\} \\ &= \ldots = \alpha(\beta\gamma).\end{aligned}$$

In the last two of these we have to use the assertion that

$$\alpha\beta = \text{mex}\{\alpha*\beta + \alpha\beta* - \alpha*\beta*\},$$

which amounts to the assertion that from $\alpha \neq \alpha*$, $\beta \neq \beta*$ we can deduce $\alpha\beta + \alpha*\beta \neq \alpha*\beta + \alpha\beta*$. But in view of the symmetry of this inequality we can suppose $\alpha > \alpha*$, $\beta > \beta*$, and the inequality is then immediate from the definition of $\alpha\beta$.

Summary. $\mathbf{On}_2$ is a commutative Ring with 1 as one.

In fact $\mathbf{On}_2$ is a Field, for we can use the analogue of our genetic construction of inverses in **No** to construct inverses in $\mathbf{On}_2$. In fact if we define $1/\alpha$ inductively by the formula

$$\beta = \frac{1}{\alpha} = \operatorname{mex}\left\{0, \frac{1 + [\alpha' - \alpha]\beta'}{\alpha'}\right\}$$

then we can mimic the proof of Chapter 1 to show that $\alpha\beta = 1$. A similar construction shows that every number in $\mathbf{On}_2$ has a square root—this time we use the inductive definition

$$\beta = \sqrt{\alpha} = \operatorname{mex}\left\{\sqrt{\alpha'}, \frac{\beta'\beta* + \alpha}{\beta' + \beta*}\right\}$$

in which β' and $\beta*$ denote options of β not both equal, which mimics Bach's definition for **No** (Chapter 1).

We shall not elaborate on these suggestions here, since in a moment we shall show that in fact $\mathbf{On}_2$ is an algebraically closed Field by a method which makes no use of these particular constructions, and enables us to locate the ordinals $1/\alpha$, $\sqrt{\alpha}$, etc., very much more easily. The results we shall prove show that each new number extends the set of previous ones in the simplest possible way, regarding addition as simpler than multiplication and division, and these as simpler than algebraic extensions which are in turn simpler than transcendental ones.

This will give us in particular a very clear picture of the field formed by the finite numbers. Thus $\{0, 1\}$ is the field $\mathbb{F}_2$ of order 2, and since this is closed under simpler operations the number 2 will define an algebraic extension, and in fact we have $2^2 = 2 + 1 = 3$, and the numbers 0, 1, 2, 3 form the field $\mathbb{F}_4$ of order 4 which is extended by the number 4 (satisfying $4^2 = 4 + 2 = 6$) to the Field $\mathbb{F}_{16}$ of order 16, and so on.

In stating our results, we follow von Neumann's convention of identifying each ordinal number with the set of all previous ones. So when we say, for instance, that 4 is a field, we mean that the set $\{0, 1, 2, 3\}$ is a field.

THE SIMPLEST EXTENSION THEOREMS

We shall frequently need to use the ordinary ordinal notions of sum, product, and power of ordinals. The ordinal sum and product are not quite the same as the maximal sum and product as used in previous chapters, but the distinction will seldom matter. We shall use [square brackets] for the ordinal operations—thus $[4 + 4] = 8$, $[4.4] = 16$, $[4^4] = 256$, whereas $4 + 4 = 0$, $4.4 = 6$, $4^4 = 4.4.4.4 = 5$.

We shall use Δ as a name for some ordinal whose arithmetic relation to earlier ordinals is currently being considered, and δ for the typical member of Δ (i.e., ordinal less than Δ).

THEOREM 39. *If Δ is not a group (under addition), then $\Delta = \alpha + \beta$, where (α, β) is any lexicographically earliest pair of numbers in Δ whose sum is not in Δ.*

Proof. Plainly $\alpha + \beta \geqslant \Delta$. But the excludents $\alpha' + \beta$, $\alpha + \beta'$ for $\alpha + \beta$ are all in Δ, so $\alpha + \beta \leqslant \Delta$.

THEOREM 40. *If Δ is a group, we have $[\Delta\alpha] + \beta = [\Delta\alpha + \beta]$ for all α, and all $\beta \in \Delta$.*

Proof. The excludents are $[\Delta\alpha' + \delta] + \beta$ and $[\Delta\alpha] + \beta'$. But since Δ is a group we can solve the equation $\delta + \beta = \bar{\delta}$ for any given $\bar{\delta} \in \Delta$, and so by induction the excludents are

$$[\Delta\alpha'] + \delta + \beta = [\Delta\alpha'] + \bar{\delta} = [\Delta\alpha' + \bar{\delta}] \text{ and } [\Delta\alpha + \beta']$$

which are precisely the numbers less than $[\Delta\alpha + \beta]$.

THEOREM 41. *If Δ is a group but not a ring, then $\Delta = \alpha\beta$, where (α, β) is any lexicographically earliest pair of numbers in Δ whose product is not in Δ.*

Proof. Plainly $\alpha\beta \geqslant \Delta$. But the excludents $\alpha'\beta + \alpha\beta' - \alpha'\beta'$ for $\alpha\beta$ are all in Δ, so $\alpha\beta \leqslant \Delta$.

THEOREM 42. *If Δ is a ring, and $\Gamma \leqslant \Delta$ is an additive subgroup all of whose non-zero elements have multiplicative inverses in Δ, then $\Delta\gamma = [\Delta\gamma]$ for all $\gamma \in \Gamma$.*

Proof. The excludents for $\Delta\gamma$ are $\Delta\gamma' + \delta(\gamma - \gamma')$. Since $\gamma - \gamma'$ is invertible in Δ, we can make $\delta(\gamma - \gamma')$ be any number $\bar{\delta}$ in Δ by choice of δ, so the typical excludent becomes

$$\Delta\gamma' + \bar{\delta} = [\Delta\gamma' + \bar{\delta}]$$

which is the typical number less than $[\Delta\gamma]$.

THEOREM 43. *If Δ is a ring but not a field, then Δ is the inverse of the earliest non-zero α in Δ which has no inverse in Δ.*

Proof. Let Γ be the largest ordinal $\leqslant \alpha$ which is a group. Then the typical excludent for $\Delta\Gamma$ is $\Delta\gamma + \delta(\Gamma - \gamma)$ $(\delta \in \Delta, \gamma \in \Gamma)$. Write $\alpha = \Gamma - \beta$.

Then for all $\gamma < \beta$, $\Gamma - \gamma$ is invertible in Δ, so that we can write $\delta(\Gamma - \gamma) = \bar{\delta}$, an arbitrary ordinal in Δ. This shows that all the numbers $[\Delta\beta' + \bar{\delta}]$ less than $\Delta\beta$ are excludents for $\Delta\Gamma$. The number $\Delta\beta = [\Delta\beta]$ is also an excludent (take $\gamma = \beta$, and $\delta = 0$). But $\Delta\beta + 1 = [\Delta\beta + 1]$ is *not* an excludent, for we should need to take $\gamma = \beta$, $\delta(\Gamma - \beta) = 1$, i.e. $\delta\alpha = 1$.

So we have $\Delta\Gamma = \Delta\beta + 1$, and so $\Delta\alpha = \Delta(\Gamma - \beta) = 1$.

THEOREM 44. *With assumptions as in Theorem 43 and its proof, we have*

$$\Delta^n\gamma_n + \Delta^{n-1}\gamma_{n-1} + \ldots + \Delta\gamma_1 + \delta = [\Delta(\Gamma^{n-1}\gamma_n + \ldots + \gamma_1) + \delta]$$

for all $n \in \omega$, and all $\gamma_0, \gamma_1, \ldots, \gamma_n \in \Gamma$, $\delta \in \Delta$.

See note on p 63.

Proof. It will suffice to prove that $\Delta^{n+1} = [\Delta\Gamma^n]$. Now the typical excludent for Δ^{n+1} has the form

$$\Delta^n(\delta_0 + \ldots + \delta_n) - \Delta^{n-1}(\delta_0\delta_1 + \ldots) + \ldots \pm \delta_0\delta_1\ldots\delta_n,$$

where the δ_i are independent variables ranging over Δ. Each of the coefficients in this polynomial is in Δ, and is either of form γ or $\Gamma + \gamma$ for some $\gamma \in \Gamma$. Using the equation $\Delta\Gamma = \Delta\beta + 1$ we can therefore reduce the polynomial to the form

$$\Delta^n\gamma_n + \ldots + \Delta\gamma_1 + \delta_0,$$

where the γ_i and δ are restricted as in the theorem. From the inductive hypothesis, we deduce that this number is less than $[\Delta\Gamma^n]$, so that $\Delta^{n+1} \leqslant [\Delta\Gamma^n]$. The opposite inequality is immediate from the inductive hypothesis.

THEOREM 45. *If Δ is a field but not algebraically closed, then Δ is a root of the lexicographically earliest polynomial having no root in Δ. [In the lexicographic order, we examine high degree coefficients first.]*

Proof. The typical excludent for Δ^n is

$$\Delta^{n-1}(\delta_1 + \ldots + \delta_n) - \Delta^{n-2}(\delta_1\delta_2 + \ldots) + \ldots \pm \delta_1\delta_2\ldots\delta_n,$$

the δ_i ranging freely over Δ.

Now if all polynomials earlier than

$$-\Delta^N + \Delta^{N-1}\alpha_{N-1} - \ldots \pm \alpha_0$$

have roots in Δ, they will all split completely into linear factors in Δ, and so we can choose n and the δ_i to show that Δ cannot be a root of any such polynomial. But if the displayed polynomial itself has no root, then every number less than

$$\Delta^{N-1}\alpha_{N-1} - \ldots \pm \alpha_0 = [\Delta^{N-1}\alpha_{N-1} - \ldots \pm \alpha_0]$$

appears as an excludent for Δ^N, but this number does not, and so we have indeed

$$\Delta^N = \Delta^{N-1}\alpha_{N-1} - \ldots \pm \alpha_0.$$

Note that we have also proved:

THEOREM 46. *With the same assumptions as in Theorem* 45 *and its proof, we have*

$$\Delta^n \delta_n + \ldots + \delta_0 = [\Delta^n \delta_n + \ldots + \delta_0]$$

for all $n < N$ *and all* $\delta_0, \ldots, \delta_n$ *in* Δ.

So it remains only to prove

THEOREM 47. *If* Δ *is an algebraically closed field, then* Δ *is transcendental over* Δ. *and we have*

$$\Delta^n \delta_n + \ldots + \delta_0 = [\Delta^n \delta_n + \ldots + \delta_0]$$

for all $n \in \omega$, *and all* $\delta_0, \ldots, \delta_n$ *in* Δ.

Proof. Any number outside an algebraically closed subfield of a larger Field *must* be transcendental over that subfield! The second part of the theorem is proved as in the previous theorems.

Summary. Each ordinal Δ extends the set of all previous ordinals in the simplest possible way, where we regard sums, products, inverses, algebraic extensions, and transcendental extensions as successively more complicated concepts.

We now turn back to the problem of identifying the first few ordinals in their role as members of $\mathbf{On}_2$.

ORDINALS BELOW THE FIRST TRANSCENDENTAL

(Some of the discussion will apply also to later ordinals.)

It is easy to see that if Δ is any group, then the next group is $[\Delta . 2]$. Hence:

THEOREM 48. *The ordinals that are groups are precisely the* 2-powers $[2^\alpha]$. *Each ordinal can be written uniquely as a finite sum of descending 2-powers, and it is the same sum in both senses.*

Proof. It follows from well-known theorems about ordinals that each ordinal has a unique expression $[2^{\alpha_0} + 2^{\alpha_1} + \ldots + 2^{\alpha_{n-1}}]$, where n is finite and $\alpha_0 > \alpha_1 > \ldots > \alpha_{n-1}$. That this is the same as $[2^{\alpha_0}] + \ldots + [2^{\alpha_{n-1}}]$ then follows from Theorem 40.

This justifies the normal rule for finding Nim-sums.

Now the ordinals below the first transcendental are algebraic over previous ones, and so by induction algebraic over the field 2 whose only elements are 0 and 1. It follows that any finite number of such ordinals generate a finite field. Each of these ordinals Δ which is itself a field defines an algebraic extension of itself. Since these extensions are taken in order of degree where possible, the first extensions will be quadratic, and then when the field is quadratically closed we shall take cubic extensions, then quintic ones, etc. [Since the Galois group of every finite field is abelian, the quadratically closed field remains quadratically closed after taking cubic extensions, etc.]

Moreover, the quadratic extensions will all be by equations of the form $x^2 + x = \alpha$, since the only lexicographically earlier quadratics are $x^2 = \alpha$, and every element of a finite field of characteristic 2 already has a square root in that field. The cubic extensions will be by cube roots, however, since the equation $x^3 = \alpha x^2 + \beta x + \gamma$ defines an extension of the field generated by α, β, γ to a larger finite field, and any finite field extension of degree 3 (and characteristic 2) can be made by a cube root, corresponding to a lexicographically earlier equation. Similar comments apply to the later extensions by fifth roots, seventh roots, etc.

THEOREM 49. *The finite numbers that are fields are the Fermat 2-powers 2, 4, 16, 256, . . . , each the [square] of the previous one. These numbers satisfy the equations*

$$2^2 = 3, 4^2 = 6, 16^2 = 24, \ldots, [2^{2^n}]^2 = [\tfrac{3}{2}.2^{2^{n+1}}].$$

The next numbers that are fields are ω, $[\omega^3]$, $[\omega^9], \ldots$, *and in the sequence*

$$2, \omega, [\omega^3], [\omega^9], \ldots$$

each term is the cube of its successor(!) *Then in the sequence*

$$4, [\omega^\omega], [\omega^{\omega 5}], [\omega^{\omega 25}], \ldots$$

each term is the fifth power of its successor, and in

$$\omega + 1, [\omega^{\omega^2}], [\omega^{\omega^2 . 7}], [\omega^{\omega^2 . 49}], \ldots$$

each term is the seventh power of its successor. In general, if p is the $(k + 1)$*st odd prime, each term in the sequence*

$$\alpha_p, [\omega^{\omega^k}], [\omega^{\omega^k \cdot p}], [\omega^{\omega^k \cdot p^2}], \ldots$$

is the p'th power of its successor, α_p *being the least number in* $[\omega^{\omega^k}]$ *with no p'th root in* $[\omega^{\omega^k}]$.

Proof. We discuss the finite number case first. It will suffice to show how the statements about 256 are deduced from those about 16. We suppose inductively that 16 is a field, and that as x varies in 16, $x^2 + x$ takes precisely the values $0, 1, \ldots, 7$ in 8. Note that when we replace x by $x + 1$, the function $x^2 + x$ is unaltered.

Then the first irreducible equation over 16 is $x^2 + x = 8$, and so we have $16^2 + 16 = 8$, whence $16^2 = 24$. Now we know that 256 is a field, with typical element $X = 16x + y$. We examine the function

$$X^2 + X = 16^2x^2 + y^2 + 16x + y = 16(x^2 + x) + (8x^2 + y^2 + y).$$

In this, $x^2 + x$ can take any value in 8, and since when we change x by 1 the expression $8x^2 + y^2 + y$ changes by 8, this expression can be made to take any value in 16 without affecting the value of $x^2 + x$. This shows that the values of $X^2 + X$ are precisely the numbers $16.8' + 16'$ in 128, which completes our inductive step.

Since ω is now known to be the quadratic closure of 2, ω must be the cube root of the smallest finite number with no finite cube root, which is 2, since any cube root of 2 has order 9, and 9 divides no number of the form $[2^{2^n} - 1]$. This statement, and the remarks before the statement of our theorem, justify the assertions about $[\omega^{3^n}]$. It is remarkable that each of these numbers is the [cube] of its predecessor, but the cube of its successor!

The numbers 2 and 3 have order 3 and so have fifth roots in ω, but 4 has order 15 (by direct calculation) and so does not, for since 25 divides no number of the form $[2^{2^n} - 1]$ there can be no finite number of order 25. The assertions about $[\omega^{\omega \cdot 5^n}]$ follow, and in a similar way we have the more general assertions about $[\omega^{\omega^k \cdot p^n}]$. We shall calculate α_7. The number 2 has [order 3 modulo 7], and 3 divides no [power of 2], so that no finite field of order $[2^{2^n}]$ can contain an element of order 7. It follows that every finite number has a finite seventh root, whence $\alpha_7 \geqslant \omega$. But the number ω lies in a cubic extension of the field 4, and so generates a field of order exactly 64. The multiplicative group of this is cyclic of order 63, and since ω has order 9, it must be a 7th power in this field. But in fact the only numbers that are 7th powers in the field are the powers of ω, which we find by direct calculation to be

$$1, \omega, \omega^2, \omega^3 = 2, \omega^4 = \omega.2, \omega^5 = \omega^2.2, \omega^6 = 3, \omega^7 = \omega.3,$$

and $\omega^8 = \omega^2.3$. Since $\omega + 1$ is not among these, it is not a seventh power in the field generated by ω and the finite numbers, and so not in any extension obtained by adjoining cube and fifth roots. So indeed $\alpha_7 = \omega + 1$.

Hendrik Lenstra has computed α_p for $p < 43$.

Observe that the theorem enables us to compute with numbers below $[\omega^{\omega^\omega}]$, using the expansion

$$\Omega^{p-1}\alpha_{p-1} + \dots + \Omega\alpha_1 + \alpha_0 = [\Omega^{p-1}\alpha_{p-1} + \dots + \Omega\alpha_1 + \alpha_0]$$

for $\Omega = [\omega^{\omega^k \cdot p^n}]$, $\alpha_0, \dots, \alpha_{p-1} < \Omega$, which follows from Theorem 46.

The theorem also tells us that $[\omega^{\omega^\omega}]$ is the algebraic closure of 2, and so is the first transcendental. Polynomials in $[\omega^{\omega^\omega}]$ with coefficients less than $[\omega^{\omega^\omega}]$ are therefore evaluated [normally], so that the next number which is a ring is $[\omega^{\omega^\omega \cdot \omega}] = [\omega^{\omega^{\omega+1}}]$.

But this ring is not a field, since $[\omega^{\omega^\omega}]$ is not invertible in $[\omega^{\omega^{\omega+1}}]$, and so $[\omega^{\omega^{\omega+1}}]$ is the inverse of $[\omega^{\omega^\omega}]$. In fact we do not see another field until we get to

$$[\omega^{\omega^{\omega^{\omega}}}] = \Omega, \text{ say.}$$

For let $t = [\omega^{\omega^{\omega}}]$, and $\alpha, \beta, \gamma, \ldots$ denote various ordinals less than t. Then since

$$[t^n\alpha + t^{n-1}\beta + \ldots + t\gamma + \delta] = t^n\alpha + t^{n-1}\beta + \ldots + t\gamma + \delta$$

we must have

$$[t^\omega] = 1/t,$$

and then we find

$$[t^{\omega+n}\alpha + \ldots + t^\omega\beta + t^m\gamma + \ldots + \delta] = \frac{\alpha}{t^{n+1}} + \ldots + \frac{\beta}{t} + \gamma t^m + \ldots + \delta$$

showing that

$$[t^{\omega+\omega}] = \frac{1}{t-1}.$$

Continuing, we find more generally that

$$[t^{\omega+\omega\alpha+n}] = \frac{1}{(t-\alpha)^{n+1}}$$

and that rational functions of t arise in lexicographic order of their partial fraction expansions

$$[\Sigma t^{\omega+\omega\alpha_i+n_i}\beta_i + \Sigma t^{m_j}\gamma_j] = \Sigma\frac{\beta_i}{(t-\alpha_i)^{n_j+1}} + \Sigma\gamma_j t^{m_j}.$$

The limit of these numbers, namely

$$[t^{\omega+\omega t}] = [t^t] = [\omega^t] = \Omega$$

must be the first algebraic extension $\Omega = \sqrt{t}$, followed by $[\Omega^2] = \sqrt[4]{t}$, $[\Omega^4] = \sqrt[8]{t}, \ldots$.

At $x = [\Omega^\omega]$ we have a perfect field, and will not need to adjoin more square roots before the next transcendental equation. In fact x satisfies

$$x^2 + x = t$$

and we must solve many such equations before the first cubic extension $\sqrt[3]{t}$, which probably happens at the next ε-number

$$\varepsilon = \left[\omega^{\omega^{\omega^{\cdot^{\cdot^{\cdot}}}}}\right],$$

Since this extension produces new quadratics to be solved, even the next cubic extension $\sqrt[3]{t-1}$ will take some time in coming.

Let us use, as sometimes customary, ε_α for the αth solution of $\alpha = [\omega^\alpha]$ (counting from $\alpha = 0$), then ζ_α for the αth solution of $\alpha = \varepsilon_\alpha$, η_α for the αth solution of $\alpha = \zeta_\alpha$, and so on through the (transfinite) Greek Alphabet. We shall use the symbol $\boxed{\alpha}$ to denote the αth letter of this Alphabet. Then we can state:

THE PROBLEM OF THE NEXT TRANSCENDENTAL

Describe in terms of ordinal arithmetic the least ordinal greater than $[\omega^{\omega^\omega}]$ which is transcendental over previous ordinals. In particular, decide where this number lies in relation to the numbers: (a) the ordinal $\boxed{\omega}_0$. (b) the least α with $\boxed{\alpha}_0 = \alpha$.

Note added in second printing

H. W. Lenstra has pointed out that the proof of Theorem 44 is incomplete. It really requires the fact that Γ is a field, which only becomes apparent later. In fact Γ is an algebraically closed field. I am also indebted to Professor Lenstra for pointing out some errors in the original version of p. 62.

Appendix to Part Zero

This is Liberty-hall, gentlemen!
Oliver Goldsmith, "She Stoops to Conquer"

In this appendix we informally discuss the formalisation of our theory, with particular regard to the nature of the inductions involved.

In Chapter 3 we gave a formal definition of the *birthday* of an arbitrary number, and we suspect that many readers would have felt happier had we described all our inductive arguments in terms of birthdays. The typical induction would then read:

"If $P(y)$ holds for all y with birthdays less than the birthday of x, then $P(x)$ holds. So by induction, $P(x)$ holds for all x."

The feeling that this sort of treatment adds to the precision of an inductive argument is much too common, and is responsible for the introduction of many irrelevancies in the literature. Thus in the case under discussion the notion of birthday is completely irrelevant, and all that is needed to justify the induction is the principle:

"If P is some proposition that holds for x whenever it holds for all x^L and x^R, then P holds universally."

We have already remarked that this was what we intended to be understood from the last sentence of our construction: "All numbers are constructed in this way."

The general inductive principle above has for its counterpart in the Zermelo–Fraenkel set theory ZF the so-called axiom of restriction, or foundation, which can be stated in the form:

"If P is some proposition that holds of a set x whenever it holds for all members of x, then P holds for every set."

Perhaps part of the prejudice against inductive arguments with arbitrary inductive variable is that this axiom is usually only stated in the peculiarly opaque form:

"Any non-empty Class X has some member disjoint from X."

It is then proved equivalent to the assertion that every set belongs to some set P_α, where these are the sets obtained by transfinite iteration of the power-set construction (P_α being the union of the power-sets of all P_β $(\beta < \alpha)$), but not to the more general inductive principle stated above. To see the latter equivalence, we need only take X as the Class of all x for which P fails.

The mention of ZF prompts a discussion of the problems of formalising our theory within ZF. Some people sense difficulties associated with the fact that an equality class of numbers is naturally a proper Class, rather than a set, and so cannot serve as an element in some other class. A slightly greater knowledge of ZF shows that this is no obstacle, and the theory may be formalised along the following lines.

We define a *game* as an ordered pair $\langle L, R\rangle$ of sets whose elements are themselves games of lower rank. (The *rank* of a set is the least α for which that set belongs to P_α.) Then we introduce the relation $\leqslant$ on games by

$$x \leqslant x' \text{ iff (no member of } L \text{ is } \geqslant x', \text{ and } x \geqslant \text{ no member of } R'),$$

where $x = \langle L, R\rangle$, and $x' = \langle L', R'\rangle$. The equivalence relation $\doteq$ is then introduced by $x \doteq y$ if and only if $x \leqslant y$ and $y \leqslant x$, and *prenumbers* are then defined inductively by the requirement that every member of $L \cup R$ should be a prenumber, and no member of $L \geqslant$ any member of R.

The fact that the equivalence classes of $\doteq$ may be proper Classes is then overcome by the standard dodge—for any x we define $[x]$ to be the set of all y *of the least possible rank* that are equivalent to x. Any set of the form $[x]$ for some prenumber x is then called a *number*.

So a number becomes a rather curiously restricted set of ordered pairs $\langle L, R\rangle$, each of which is of course a set according to the Kuratowski definition $\langle L, R\rangle = \{\{L\}, \{L, R\}\}$.

Another, and technically simpler, approach makes use of the sign-expansions introduced in Chapter 3. We define a number to be its sign-expansion, which is of course a function from some ordinal α to the set $\{+, -\}$. We then define order-relations in terms of these expansions by the rules in Chapter 3, and *define* $\{L \mid R\}$ to be the simplest (i.e. *shortest*) number greater than every member of L and less than every member of R. We then define the arithmetic operations by the formulae in Chapter 0.

In this simpler formalisation, a number is still a pretty complicated thing, namely a certain function in ZF, which is of course a certain set of Kuratowskian ordered pairs. The first members of these ordered pairs will be ordinals in the sense of von Neumann, and the second members chosen from the particular two-element set we take to represent $\{+, -\}$.

The curiously complicated nature of these constructions tells us more about the nature of formalisations within ZF than about our system of numbers, and it is partly for this reason that we did not present any such formalised theory in this book. But the main reason was that we regard it as almost self-evident that our theory is as consistent as ZF, and that formalisation in ZF destroys a lot of its symmetry. Plainly the proper set theory in which to perform a formalisation would be one with two kinds of membership, and would in fact be very like the abstract theory of games that underlies the next part of this book.

It seems to us, however, that mathematics has now reached the stage where formalisation within some particular axiomatic set theory is irrelevant, even for foundational studies. It should be possible to specify conditions on a mathematical theory which would suffice for embeddability within ZF (supplemented by additional axioms of infinity if necessary), but which do not otherwise restrict the possible constructions in that theory. Of course the conditions would apply to ZF itself, and to other possible theories that have been proposed as suitable foundations for mathematics (certain theories of categories, etc.), but would not restrict us to any particular theory. This appendix is in fact a cry for a Mathematicians' Liberation Movement!

Among the permissible kinds of construction we should have:

(i) Objects may be created from earlier objects in any reasonably constructive fashion.

(ii) Equality among the created objects can be any desired equivalence relation.

In particular, set theory would be such a theory, sets being constructed from earlier ones by processes corresponding to the usual axioms, and the equality relation being that of having the same members. But we could also, for instance, freely create a new object (x, y) and call it the ordered pair of x and y. We could also create an ordered pair $[x, y]$ different from (x, y) but co-existing with it, and neither of these need have any relation to the set $\{\{x\}, \{x, y\}\}$. If instead we wanted to make (x, y) into an unordered pair, we could define equality by means of the equivalence relation $(x, y) = (z, t)$ if and only if $x = z, y = t$ *or* $x = t, y = z$.

I hope it is clear that this proposal is not of any particular theory as an alternative to ZF (such as a theory of categories, or of the numbers or games considered in this book). What is proposed is instead that we give ourselves the freedom to create arbitrary mathematical theories of these kinds, but prove a metatheorem which ensures once and for all that any such theory could be formalised in terms of any of the standard foundational theories.

The situation is analogous to the theory of vector spaces. Once upon a time these were collections of n-tuples of numbers, and the interesting theorems were those that remained invariant under linear transformations of these numbers. Now even the initial definitions are invariant, and vector spaces are defined by axioms rather than as particular objects. However, it is proved that every vector space has a base, so that the new theory is much the same as the old. But now no particular base is distinguished, and usually arguments which use particular bases are cumbrous and inelegant compared to arguments directly in terms of the axioms.

We believe that mathematics itself can be founded in an invariant way, which would be equivalent to, but would not involve, formalisation within some theory like ZF. No particular axiomatic theory like ZF would be needed, and indeed attempts to force arbitrary theories into a single formal strait-jacket will probably continue to produce unnecessarily cumbrous and inelegant contortions.

For those who doubt the possibility of such a programme, it might be worthwhile to note that certainly principles (i) and (ii) of our Mathematicians' Lib movement can be expressed directly in terms of the predicate calculus without any mention of sets (for instance), and it can be shown that any theory satisfying the corresponding restrictions can be formalised in ZF together with sufficiently many axioms of infinity.

Finally, we note that we have adopted the modern habit of identifying ZF (which properly has only sets) with the equiconsistent theory NBG (which has proper Classes as well) in this appendix and elsewhere. The classification of objects as Big and small is not peculiar to this theory, but appears in many foundational theories, and also in our formalised versions of principles (i) and (ii).

FIRST PART

. . . AND GAMES

But leave the Wise to wrangle, and with me
The Quarrel of the Universe let be:
And, in some corner of the Hubbub coucht,
Make Game of that which makes as much of Thee

The Rubaiyat of Omar Khayyam

CHAPTER 7

Playing Several Games at Once

For when the One Great Scorer comes
to write against your name,
He marks—not that you won or lost—
but how you played the game.

Grantland Rice,
Alumnus Football

The games we shall consider are in spirit closer to Chess than to Football. We imagine them played, on some kind of board perhaps, between two players whose usual names are *Left* and *Right.* [Aliases (respectively) Black and White, Vertical and Horizontal, Arthur and Bertha.] Our own sympathies are usually with Left.

The games these people play have *positions*, and in any position P, there are rules which restrict Left to move to any one of certain positions (typically P^L) called the *Left options* of P, while Right may similarly move only to certain positions (typically P^R) called the *Right options* of P. Since we are interested only in the abstract structure of games, we can regard any position P as being completely determined by its Left and Right options, and so we shall write $P = \{P^L \mid P^R\}$.

Thus if in some game there is a position P from which Left may move to any one of certain positions A, B, C (only), while Right may move only to the position D, then we write $P = \{\text{A, B, C} \mid \text{D}\}$.

A game obviously *ends* when the player who is called upon to move finds himself unable to do so. So for instance the position $\{ \mid \text{U, V, W, X}\}$, with Left about to move, obviously corresponds to an ended game. Except in Chapters 12 and 14, we adopt the *normal play* convention, according to which *a player who is unable to move when called upon to do so is the loser.* This is obviously a natural convention, for since we normally consider ourselves as losing when we cannot find any good move, we should obviously lose when we cannot find any move at all!

Our players Left and Right are usually unwilling to play games that are capable of going on forever (they are both busy men, with heavy political responsibilities). So except for a moment in Chapter 11, we adopt the convention that in no game is there an infinite sequence of positions each of which is an option of its predecessor. [Including *in particular* the case when these options are alternately Left and Right.]

Each game G has its own proper *starting position*, the position from which we usually start to play. But for any position P of G we can obviously obtain a shortened game by starting instead at P. We find it handy to identify this game with P, so that in particular every game G will automatically be identified with its starting position.

It follows from these conventions that games can be represented by *trees*, the positions being represented by nodes (the initial position being the lowest node, or *root*), and the legal moves by branches. We shall always draw these trees so that the moves for Left are represented by leftwards slanting branches, and those for Right by rightwards slanting ones.

EXAMPLES OF SIMPLE GAMES

In Fig. 4 we draw these trees for the four simplest games (born on days 0 and 1),

$0 = \{\,|\,\}$ $\quad 1 = \{0\,|\,\}$ $\quad -1 = \{\,|\,0\}$ $\quad * = \{0\,|\,0\}$

FIG. 4. The simplest games.

The simplest game of all is the *Endgame*, 0. I courteously offer you the first move in this game, and call upon you to make it. You lose, of course, because 0 is defined as the game in which it is never legal to make a move.

In the game $1 = \{0\,|\,\}$, there is a legal move for Left, which ends the game, but at no time is there any legal move for Right. If I play Left, and you Right, and you have first move again (only fair, as you lost the previous game) you will lose again, being unable to move even from the initial position. To demonstrate my skill, I shall now start from the same position, make my legal move to 0, and call upon you to make yours.

Of course you are now beginning to suspect that Left always wins, so for our next game, -1, you may play as Left and I as Right! For the last of our examples, the new game $* = \{0\,|\,0\}$, you may play whichever role you wish, provided that for this privilege you allow me to play first.

We summarise your probable conclusions:

In the game 0, there is a winning strategy for the second player

In the game 1, there is a winning strategy for Left (whoever starts)

In the game -1, there is a winning strategy for Right; and, finally,

In the game $*$, there is a winning strategy for the first player to move.

In general, we introduce corresponding notations:

$G > 0$ (G is *positive*) if there is a winning strategy for Left

$G < 0$ (G is *negative*) if there is a winning strategy for Right

$G = 0$ (G is *zero*) if there is a winning strategy for the second player,

$G \parallel 0$ (G is *fuzzy*) if there is one for the first player.

We shall also combine these symbols:

$G \geqslant 0$ means $G > 0$ *or* $G = 0$; $G \leqslant 0$ means $G < 0$ *or* $G = 0$;

$G \mathrel{|\triangleright} 0$ means $G > 0$ *or* $G \parallel 0$; $G \mathrel{\triangleleft|} 0$ means $G < 0$ *or* $G \parallel 0$.

Thus $G \geqslant 0$ means that supposing *Right* starts, there is a winning strategy for Left, while $G \mathrel{|\triangleright} 0$ means that there is a winning strategy for Left if *Left* starts. In slightly less formal terms, justified by Theorem 50, we can say that $G \geqslant 0$ if there is *no* winning first move for Right (the start of a winning strategy for him), while $G \mathrel{|\triangleright} 0$ means that there *is* a winning first move for Left.

THEOREM 50. *Each game G belongs to one of the outcome classes above.*

Proof. This is equivalent to the assertion that for each game G, we have either $G \geqslant 0$ or $G \mathrel{\triangleleft|} 0$, and either $G \leqslant 0$ or $G \mathrel{|\triangleright} 0$. Suppose that this is true of all G^L, G^R. Then if any $G^L \geqslant 0$, Left can win by first moving to this G^L, and then following with his strategy for this G^L, Right starting. If not, we have each $G^L \mathrel{\triangleleft|} 0$, and Right has a winning strategy in G, Left starting. He just sits back and waits until Left has moved to some G^L, and then applies his winning strategy (Right starting) in that G^L.

THE NEGATIVE OF A GAME

Since the legal moves for the two players are not necessarily the same, we may obtain a distinct game by reversing the roles of Left and Right throughout G. The game so obtained we call the *negative* of G. Inductively, it is the game $-G$ defined by the equation

$$-G = \{-G^R \mid -G^L\}.$$

Obviously, negation interchanges positive and negative games, while the negative of a zero or fuzzy game is another game of the same type.

SIMULTANEOUS DISPLAYS. SUMS OF GAMES

Left and Right are given to playing simultaneous displays of games against each other, in the following manner. Each game is placed on a table, and when it is Left's turn to move, he selects one of the component games, and makes any move legal for Left in that game. Then Right selects some component game (possibly the same as that used by Left, possibly not), and makes a move legal for Right in that game. The game continues in this way until some player is unable to move in any of the components, when of course that player loses, according to the normal play convention.

When games G and H are played as a simultaneous display in this manner, we refer to the compound game as the *disjunctive sum* $G + H$ of the two games. Most of the rest of this book is concerned with such disjunctive sums—which we therefore simply call *sums*—but in Chapter 14 we shall consider some other kinds of simultaneous display, which will lead to other operations on games.

HOW SUMS HAPPEN—A GAME WITH DOMINOES

In fact it often happens in some real-life game that a position breaks up into a disjunctive sum, because it is obvious for some reason that moves made in one part of the position will not affect the other parts. Consider for example the following game with dominoes, suggested by Göran Andersson.

On a rectangular board ruled into squares, the players alternately place dominoes which cover two adjacent squares, Left being required to place his dominoes vertically, Right horizontally. The dominoes must not overlap, and the last player able to move is the winner.

After a time, the vacant spaces left on the board are usually in several separated regions, and the game becomes a sum of smaller games one for each region. We analyse the simplest possibilities.

A region □ contains no move for either player, and so is abstractly the game $\{\,|\,\} = 0$. Such regions can be neglected.

A region ⊟ or ⊟ has just one move for Left (to 0), but none for Right. Its value is therefore $\{0\,|\,\} = 1$, and indeed it confers an advantage of just one move upon Left. Similarly the region □□□□ is -2, since it has no move for Left, but moves for Right to 0 and -1, and we recall $\{\,|\,0, -1\} = -2$.

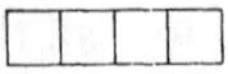

In general, if a position has no move for Right at any time, and at most n successive moves for Left, its value is n, and the value will be $-n$ if we reverse the roles of Left and Right here.

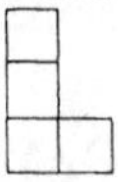

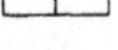

The region ⊞ is more interesting. Left has one (stupid) move to □□ $= -1$ and another (more sensible) move to □ + □ $= 0$, whereas Right has only one move to ⊟ $= 1$. So the value should be $\{0, -1\,|\,1\}$, which the diligent reader of the zeroth part of this book will recognise as $\frac{1}{2}$. And there is indeed a definite sense in which this region represents an advantage of exactly one half of a move to Left!

Values other than numbers can occur in this domino game. The region 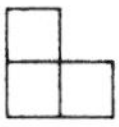 has value $\{0 \mid 0\} = *$, since either player can move to $= 0$ (only), while the region 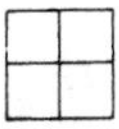 has value $\{1 \mid -1\}$ since Left moves to 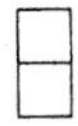 $= 1$, and Right by symmetry to -1.

The dominoes position with regions 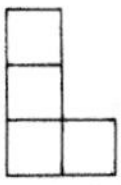 ,  , 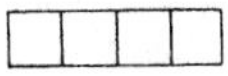 (only) has the value $\frac{1}{2} + 1 - 2 = -\frac{1}{2}$. Since this is *negative*, Right is half-a-move ahead, and can win the game, no matter who starts.

SUMS OF SIMPLE GAMES

Since it is never legal to move in 0, the game $G + 0$ is essentially the same as G, and we write $G + 0 = G$.

The game $1 + 1$. From the sum $1 + 1$, Left can move to $1 + 0$ or $0 + 1$, both essentially the same as 1. Since Right can never move, we have $1 + 1 = \{1, 1 \mid \}$, and since Left's two moves are essentially the same, we can simplify this further to $1 + 1 = \{1 \mid \}$. This game we call 2. It is a positive game, since Left has moves but Right has not.

The game $1 - 1$. We write $1 - 1$ for the sum $1 + (-1)$. In this, Left can only move to $0 + -1 = -1$ (which is a win for Right), and Right can only move to $1 + 0$, a win for Left. So neither player will really want to move, and the game is a zero game. In symbols, we have $1 - 1 \equiv \{-1 \mid 1\} = 0$.

The game $* + *$. In a similar way, $* + * \equiv \{* \mid *\}$, which, since $*$ is a win for the first player, is a second player win. So we have $* + * = 0$.

What do these equalities mean?

There is a famous story of the little girl who played a kind of simultaneous display against two Chess Grandmasters (surely a Big concept!). How was it that she managed to win one of the games? Anne-Louise played Black against Spassky, White against Fischer. Spassky moved first, and Anne-Louise just copied his move as the first move of her game against Fischer, then copied Fischer's reply as her own reply to Spassky's first move, and so on.

THEOREM 51. *$G - G$ is always a zero game.*

Proof. The moves legal for one player in G become legal for his opponent in $-G$, and vice versa. So the second player can win $G - G$ by always mimicking her opponent's previous move—if Left moves to G^L in G, Right (as second player) can move to $-G^L$ in $-G$. If she plays in this way, the second player will never be lost for a move in $G - G$.

In a similar way, we can prove:

THEOREM 52. *From $G \geqslant 0$ and $H \geqslant 0$, we can deduce $G + H \geqslant 0$.*

Proof. The suppositions tell us that if Right starts, Left can win each of G and H. But he can then win $G + H$ by always replying in the component Right moves in, and making the winning reply in this component. In this way, Left cannot be lost for a move in G or H, and so will win the sum.

THEOREM 53. *If H is a zero game, then $G + H$ has the same outcome as G.*

Proof. This can be made to follow from the previous theorem, but we give it a separate proof. Play $G + H$, in exactly the same way as you would in G, never moving in the H component except to reply to an immediately previous move of your opponent in that game. This rule converts a winning strategy for you in G to one for you in $G + H$, it being understood that the same player starts in both cases.

THEOREM 54. *If $H - K$ is a zero game, then the games $G + H$ and $G + K$ have always the same outcome.*

Proof. $G + K$ has the same outcome as $(G + K) + (H - K)$, by Theorem 53. But this can be written as $(G + H) + (K - K)$, which has the same outcome as $G + H$, since $K - K$ is a zero game.

Now our aim in this book is to find out who wins sums of various games, so that if $H - K$ is a zero game, it will not matter if we replace H by K. So in this case, we shall say that H is *equal* to K, and write $H = K$. We shall not usually distinguish between equal games, and so when we speak of the game 0, we mean to refer also to the games $1 - 1$, $* + *$, and so on. On occasions when it is necessary to make these distinctions, we speak of the *form* of a game (meaning some particular game, regarded as distinct from its equals) and the *value* of a game (G and H having the same value when $G = H$).

SOME MORE GAMES

The game $\frac{1}{2}$. We define $\frac{1}{2} = \{0 \mid 1\}$, and verify the equality $\frac{1}{2} + \frac{1}{2} = 1$. In Fig. 5 we have drawn the components of the game $\frac{1}{2} + \frac{1}{2} - 1$, with letters for the names of various positions.

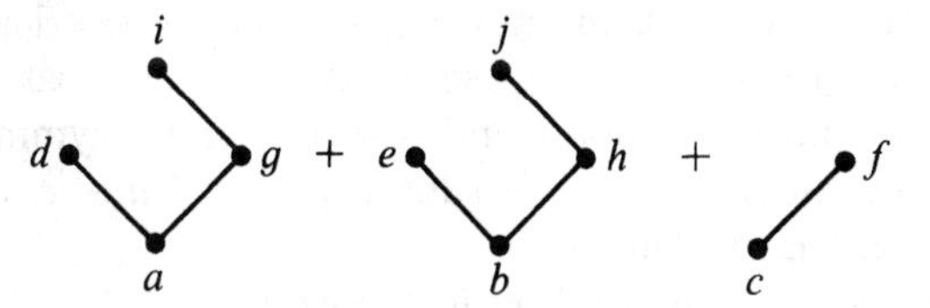

FIG. 5. Strategic proof that $\frac{1}{2} + \frac{1}{2} = 1$.

Initially, we are at the position (a, b, c). We consider first what happens if Left starts. He might as well move from a to d, to which Right replies by the move from b to h, then Left can only move from h to j, and Right makes the last move from c to f and wins.

If Right moves from b to h, Left can reply with a to d, and then wins with h to j as his reply to Right's only move c to f. If instead Right makes the move c to f, Left can reply a to d, then we have b to h for Right, followed by the winning move h to j. (Note that in all cases we have the same 4 moves $a \to d$, $b \to h$, $h \to j$, $c \to f$. This phenomenon often happens.)

Exercise. Taking $\frac{1}{4}$ as $\{0 \mid \frac{1}{2}\}$ and $\frac{3}{4}$ as $\{\frac{1}{2} \mid 1\}$, give a strategic discussion of the equality $\frac{1}{2} + \frac{1}{4} = \frac{3}{4}$.

The game $\uparrow$. The game $\{0 \mid *\}$ is common enough to deserve a special name, so we call it *up*, and give it the special symbol $\uparrow$. Its negative $\{* \mid 0\}$—note that $*$ is its own negative, like 0—is called *down* and given the symbol $\downarrow$. Since Left wins with the first or second move, $\uparrow$ is a positive game. It is the value of the position in our domino game. In Fig. 6 we illustrate the remarkable equality

$$\{0 \mid \uparrow\} = \uparrow + \uparrow + *.$$

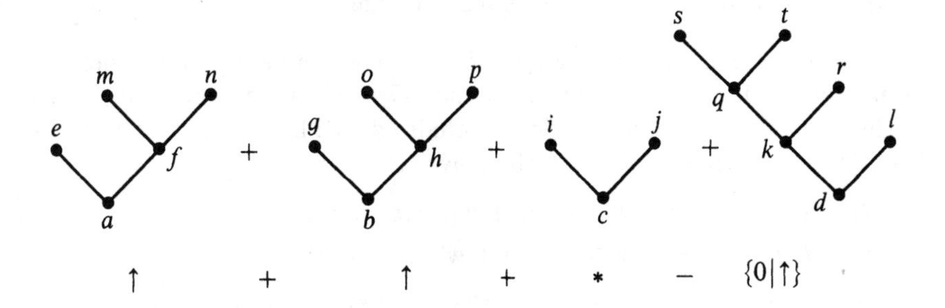

FIG. 6. The upstart equality.

In the illustrated position, the moves $a \to f$ and $d \to k$ lead collectively to the zero position $* + \uparrow + * + \downarrow$, so we can use either as a reply to the other, and then mimic our opponent's moves. So by symmetry we need only consider the moves $c \to j$, $d \to l$ for Right, and $a \to e$, $c \to i$ for Left, showing that each has its counter.

Now the moves: $c \to j$, $d \to k$ lead to a position $\uparrow + \uparrow + \downarrow = \uparrow > 0$, and $d \to l$, $c \to i$ lead to $\uparrow + \uparrow > 0$, so that $c \to j$ and $d \to l$ are bad moves for Right. Similarly, after $a \to e$, $b \to h$, we have a position $* + * + d = d$, and Right wins d, the moves being $d \to k$, $k \to r$. In the final case, Right replies to $c \to i$ with $a \to f$, and then follows one of ($f \to m$, $b \to h$), ($b \to g$, $f \to n$), and ($d \to k$, $f \to n$) and an easy win for Right in each case. So indeed we have $\uparrow + \uparrow + * = \{0 \mid \uparrow\}$.

We close this introductory chapter with the details of a more formal approach, for those who might prefer it.

Construction. If L and R are any two sets of games, there is a game $\{L \mid R\}$. All games are constructed in this way.

Convention. If $G = \{L \mid R\}$, we write G^L for the typical element of L, G^R for the typical element of R, and refer to these (respectively) as the Left and Right options of G. Then the *legal moves* in G are, for Left, from G to G^L, and for Right, from G to G^R, and we write $G = \{G^L \mid G^R\}$.

Definition of $G \geqslant H$, etc.
$G \geqslant H$ iff (no $G^R \leqslant H$ and $G \leqslant$ no H^L). $G \leqslant H$ iff $H \geqslant G$. $G \| H$ iff neither. $G \mathrel{|\!\!\triangleright} H$ iff $G \nleqslant H$; $G \mathrel{\triangleleft\!\!|} H$ iff $G \ngeqslant H$; $G < H$, $G > H$, $G =$ H, as usual.

Definition of $G + H$.

$$G + H = \{G^L + H, G + H^L \mid G^R + H, G + H^R\}$$

Definition of $-G$.

$$-G = \{-G^R \mid -G^L\}.$$

Then we have all the statements of the following.

Summary. The Class **Pg** of all *Partizan Games* forms a partially ordered group under addition, with 0 as zero and $-G$ as negative, when considered modulo equality. This Group strictly includes the additive Group of all numbers. The order-relation is that defined by

$G > H$ iff $G - H$ is won by Left, whoever starts

$G < H$ iff $G - H$ is won by Right, whoever starts

$G = H$ iff $G - H$ is won by the second player to move, and

$G \| H$ iff $G - H$ is won by the first player to move.

The relation $G \| H$ is the relation of incomparability for this order, meaning that we have no one of $G = H$, $G > H$, $G < H$. We say then that G and H are *confused*, or that G *is fuzzy against* H.

Formal proofs of these statements from these definitions are to be found in the zeroth part of this book where in some places we were careful to word our proofs so as to include more general games, although we were then primarily interested in numbers. Informal proofs and explanations in terms of strategies have been given in this chapter.

However, there is one point that calls for special notice. The phrase "all games are constructed in this way" justifies the proving of theorems by induction over games. Thus if for all G we can deduce that P holds at G provided it holds at all options of G, then P holds for all games. The following argument shows that this is equivalent to our requirement that there be no infinite sequence of games each an option of its predecessor.

If such a property P does not hold for some game $G = G_0$, then it must also fail for some option G_1 of G_0, and then for some option G_2 of G_1, and so on. So unless P holds for all games, we obtain an infinite option-sequence. [This proof uses the axiom of choice.]

SOME INFINITE GAMES

At first sight it might be thought that the previous discussion makes all games finite. But the game $\omega = \{0, 1, 2, 3, \ldots |\}$ has infinitely many positions, and yet is a perfectly good game, if a little biassed in favour of Left. For since after the first move, we reach some finite game $n = \{0, 1, 2, \ldots, n-1\}$, which lasts at most n moves, there can be no infinite option-sequence in ω. But of course we can give no fixed estimate, before choosing the first option, for the length of an option-sequence. The tree of ω is sketched in Fig. 7.

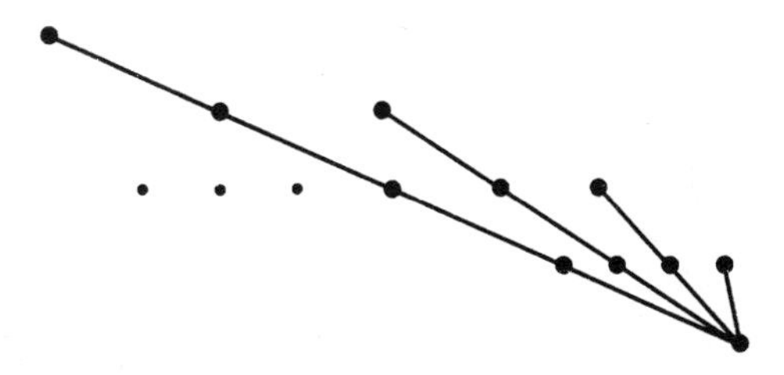

FIG. 7. The tree of ω.

MY DAD HAS MORE MONEY THAN YOURS

In this game, the players alternately name sums of money (for just two moves), and the player who names the larger amount is the winner. The game is essentially the same as

$$\omega - \omega = \{0 - \omega, 1 - \omega, \ldots, n - \omega, \ldots \mid \omega - 0, \omega - 1, \ldots, \omega - n, \ldots\},$$

whose tree is rather complicated, though the complication is irrelevant in play. As childhood experience shows, there is not much point in starting first at this game. This observation is equivalent to the equality

$$\omega - \omega = 0.$$

The theory of games developed in the rest of this book is a grand generalization of the earlier theory found independently by Sprague and Grundy for *impartial games*—those in which both players have the same legal moves. In the first edition of this book the term "unimpartial" was used for the wider class of games obtained by dropping this condition—we now adopt the nicer word "partizan" that was introduced in *Winning Ways*.

CHAPTER 8

Some Games are Already Numbers

"Reeling and Writhing, of course, to begin with," the Mock Turtle replied; "And then the different branches of Arithmetic— Ambition, Distraction, Uglification, and Derision."

Lewis Carroll, "Alice in Wonderland".

In this chapter we consider several games in which the values of all, or almost all, the positions are already numbers. For such a game we shall obtain a complete theory as soon as we can give some rule for calculating the number which is the value of any particular position. We shall not always be able to do this, even when we can quite easily prove that all the values are numbers.

The diligent reader of the zeroth part of this book will already know quite a lot about numbers. But for the benefit of certain other readers, we summarise some of the more basic information here.

There is a notion of *simplicity* for numbers, which we can if we like define as follows. [This is not quite the same as the notion used in the zeroth part, but the differences are inessential.]

The number 0 is the simplest possible number, followed by the numbers 1 and -1, then 2 and -2, 3 and -3, etc., and so on through all the integers. Next come all rationals with denominator 2, followed by those with denominator 4 (*not* 3), then those with denominator 8, and so on through the dyadic rationals. After these come all remaining real numbers at once, including $\frac{1}{3}$, $\sqrt{2}$, and π as examples.

For the extensions to other numbers, see the tree in Chapter 0, the discussion in Chapter 3, and some of the remarks in the appendix to the zeroth part. In this part of the book we shall mostly talk only about ordinary real numbers, and the above discussions should be enough, but for the occasional comments about other surreal numbers we shall suppose that the reader is familiar with the zeroth part.

The most important game-theoretical property of numbers is that given by the *simplicity rule*: if all the options G^L and G^R of some game G are known to be numbers, and each G^L strictly less than each G^R, then G is itself a number, namely the simplest number x *greater than* every G^L and *less than* every G^R. (Theorem 11, Chapter 2.)

CONTORTED FRACTIONS

This game is actually played with numbers, so that it is not surprising that numbers arise in its solution. However, the complete theory is rather curious.

The typical position has a number of real numbers in boxes, and the typical legal move is to alter just one of these numbers. The number replacing a given one must have strictly smaller denominator, or, if the given number was already an integer, be an integer strictly smaller in absolute value. Irrational numbers are counted as having infinite denominator. Such a replacement will be legal for Left only if it *decreases* the number, legal for Right only if it *increases* it.

Thus from the position $\boxed{\frac{2}{5}}$ Left can move to the positions $\boxed{x}$ with $x = \frac{1}{3}, \frac{1}{4}, -\frac{1}{2}, 0, -2$, etc., since all these are less than $\frac{2}{5}$ and have denominator smaller than 5, and Right can similarly move to $\boxed{x}$ with $x = \frac{1}{2}, \frac{2}{3}, \frac{3}{4}, 1, 17\frac{1}{4}$, etc. But in general Left will prefer to keep the numbers as large as possible, while Right will wish to make them small, so that in fact Left will choose $x = \frac{1}{3}$ and Right $x = \frac{1}{2}$, if they play wisely. In symbolic terms, this means that we have the equation

$$\boxed{\tfrac{2}{5}} = \{\boxed{\tfrac{1}{3}} \,|\, \boxed{\tfrac{1}{2}}\}.$$

So it is fairly easy to see that what has happened in this game is that we have imposed a distorted notion of simplicity, under which $\frac{1}{3}$ is counted as simpler than $\frac{1}{4}$ because it has smaller denominator. Proceeding in order of this new kind of simplicity, we obtain the table

$$x = \ldots -1 \;\; -\tfrac{1}{2} \;\; 0 \;\; \tfrac{1}{4} \; \tfrac{1}{3} \; \tfrac{2}{5} \; \tfrac{1}{2} \; \tfrac{3}{5} \; \tfrac{2}{3} \; \tfrac{3}{4} \;\; 1 \;\; 1\tfrac{1}{2} \;\; 2 \ldots$$

$$\boxed{x} = \ldots -1 \;\; -\tfrac{1}{2} \;\; 0 \;\; \tfrac{1}{8} \; \tfrac{1}{4} \; \tfrac{3}{8} \; \tfrac{1}{2} \; \tfrac{5}{8} \; \tfrac{3}{4} \; \tfrac{7}{8} \;\; 1 \;\; 1\tfrac{1}{2} \;\; 2 \ldots$$

in which arbitrary fractions on the top line correspond to dyadic ones on the bottom line, in the respective orders of simplicity.

The well-known rule for Farey fractions tells us how to find new entries successively—if a/b and c/d are at some time adjacent in the top line, then the next number to insert between them is $(a + c)/(b + d)$, and so this number will yield the mean of the two numbers corresponding to a/b and c/d in the bottom line. (This only happens if $bc - ad = 1$.) Thus we have the equation $\boxed{\frac{4}{7}} = \frac{9}{16}$, operating in this way on the adjacent numbers $\frac{1}{2}$ and $\frac{3}{5}$ from the top line.

The general solution requires some of the theory of continued fractions, and since this is no part of our business here, we shall simply quote the answer. The proof involves also Berlekamp's rule for interpreting sign-expansions (Chapter 3).

Each rational number x can be expanded as a simple continued fraction in two closely related ways:

$$x = a + \frac{1}{b+}\frac{1}{c+}\cdots+\frac{1}{n+1} = a + \frac{1}{b+}\frac{1}{c+}\cdots+\frac{1}{n+}\frac{1}{1},$$

in view of the equation

$$\frac{1}{n+(1/1)} = \frac{1}{n+1}.$$

We obtain from this continued fraction expansion for x the dyadic rational value for $\boxed{x}$ as follows.

Write down the integer a, with its sign, as the integral part of $\boxed{x}$. For the fractional part, we have the binary expansion $\cdot\, 0^{b-1}1^c0^d\ldots$, where we choose the particular representation so that this ends in 1. In other words, we read the partial quotients $b, c, \ldots$ as alternate numbers of 0s and 1s, except that the first 0 is replaced by the binary point.

Thus

$$2\tfrac{13}{35} = 2 + \frac{1}{2+}\frac{1}{1+}\frac{1}{2+}\frac{1}{4},$$

and so we have

$$\boxed{2\tfrac{13}{35}} = 2{\cdot}01001111 = 2\tfrac{79}{256}.$$

(The alternative form

$$2 + \frac{1}{2+}\frac{1}{1+}\frac{1}{2+}\frac{1}{3+}\frac{1}{1}$$

would yield a binary expansion ending in 0, and so is discarded.) Of course the numbers before the binary point will usually be written in decimal, so that we have a curiously mixed notation here!

For irrational x, we obtain an infinite continued fraction, and exactly the same rule works, except that we have no worries about double representation. Thus for

$$x = 1 + \frac{1}{1+}\frac{1}{1+}\cdots = 1 + \frac{1}{x},$$

we have the binary expansion $1{\cdot}101010\ldots = 1\tfrac{2}{3}$. Since this x is the positive root of the equation $x^2 = x + 1$, we have the mystic equation

$$\boxed{\frac{1 + \sqrt{5}}{2}} = \tfrac{5}{3}.$$

The function here called $\boxed{x}$ is traditionally called "Minkowski's Question-Mark Function," and has interesting analytic properties. Its graph is shown in Fig. 8.

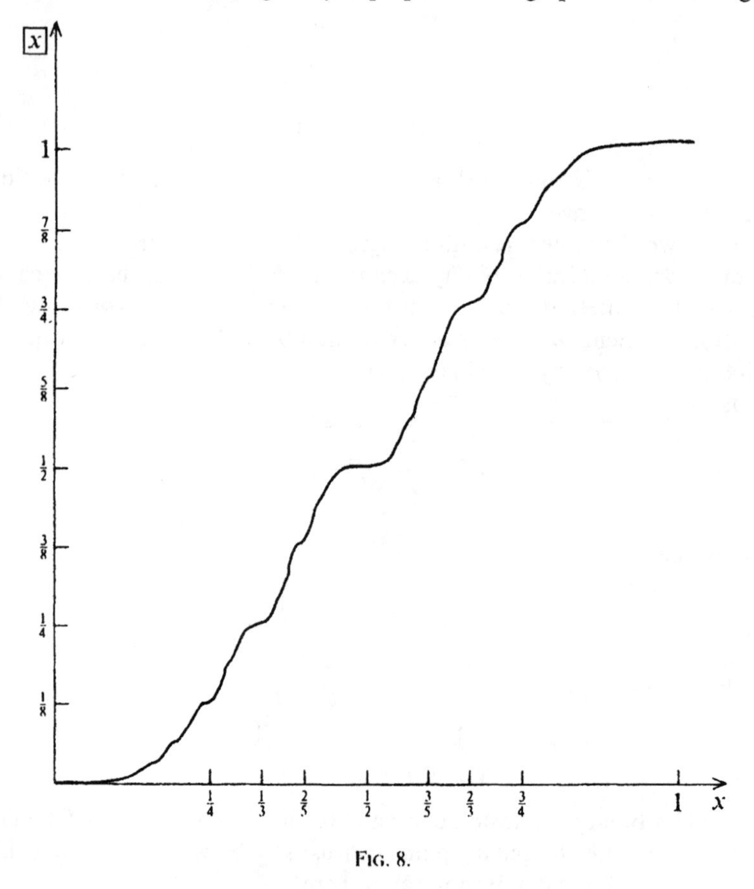

FIG. 8.

Suppose we have the position

$$\boxed{x} + \boxed{x} + \boxed{x} + \boxed{x} + \boxed{x}$$

but that Right is allowed to pass just once during the game, at any time he chooses. For what real number x is this a fair game?

The allowance for Right is equivalent to adding an extra component -1, and so we must solve the equation $\boxed{x} = \frac{1}{5}$. Now the number $\frac{1}{5}$ has the binary expansion 0·00110011001100..., and so the required x is the number represented by the continued fraction

$$x = \frac{1}{3+}\,\frac{1}{2+}\,\frac{1}{2+}\cdots.$$

Now writing t for the number

$$1 + \frac{1}{2+}\,\frac{1}{2+}\cdots$$

we find that

$$t = 1 + \frac{1}{1+t},$$

and so $t^2 = 2$, whence $t = \sqrt{2}$ since t is obviously positive, and this gives us the surprising answer

$$x = \frac{1}{2+\sqrt{2}}.$$

Problems. Solve the equations

$$\boxed{\sqrt{2}} + \boxed{\sqrt{3}} = \boxed{A} \qquad \left(A = \tfrac{13}{3} - \sqrt{\tfrac{26}{27}}\right)$$

$$\boxed{\sqrt{5}} - \boxed{\sqrt{2}} = \boxed{B} \qquad \left(B = 1 - \sqrt{\tfrac{2}{15}}\right)$$

$$4\boxed{e} = \boxed{C} \qquad \left(C = \frac{25e - 63}{2e - 5}\right)$$

$$\boxed{\pi} + \boxed{\tfrac{1}{2}} = \boxed{D} \qquad \left(D = \frac{37 - 10\pi}{11 - 3\pi}\right)$$

$$\boxed{\pi} - \boxed{\tfrac{1}{100}} = \boxed{E} \qquad \left(E = \frac{240585707\pi - 755822109}{76580827\pi - 240585706}\right)$$

We illustrate with the last equation (none of the others requires much calculation). The continued fraction for π is

$$\pi = 3 + \frac{1}{7+}\,\frac{1}{15+}\,\frac{1}{1+}\,\frac{1}{292+}\cdots$$

which we write as

$$3 + \frac{1}{7+}\,\frac{1}{15+}\,\frac{1}{1+}\,\frac{1}{77+x}$$

for a reason that will soon become apparent. So the expansion of $\boxed{\pi}$ will be

$$\underbrace{3{\cdot}000000}_{7}\ \underbrace{1111111111111110}_{15}\ \underbrace{1111111111\ldots111\underline{1}}_{1\ (76\ +\ 1}\underbrace{11111\ldots1110}_{+\ 215)}\ldots,$$

and

0·000. 0001

is the corresponding expansion of $\boxed{\frac{1}{100}}$. We conclude that E must be the number

$$3+\frac{1}{7+}\,\frac{1}{15+}\,\frac{1}{1+}\,\frac{1}{76+}\,\frac{1}{1+}\,\frac{1}{215+}\ldots,$$

or more simply

$$E=3+\frac{1}{7+}\,\frac{1}{15+}\,\frac{1}{1+}\,\frac{1}{76+}\,\frac{1}{1+}\,\frac{1}{x}.$$

Eliminating x we find the displayed answer. The calculations would have been *much* harder if we had not the good rational approximation $\pi \doteqdot \frac{355}{113}$!

HACKENBUSH RESTRAINED

In this game, the appearance of the numbers is less expected, but they also appear less curiously. The game has analogues and generalisations which will be considered in other chapters. This variety of Hackenbush is played on a *picture*, consisting of black edges (▌) and white edges (▯) joining nodes. It is required that each node be connected via a chain of edges to a certain dotted line called the *ground* (sometimes also called the *ceiling*, or the *walls*). Two nodes may be joined by more than one edge, and it may happen that some edge joins a node to itself. See Fig. 9.

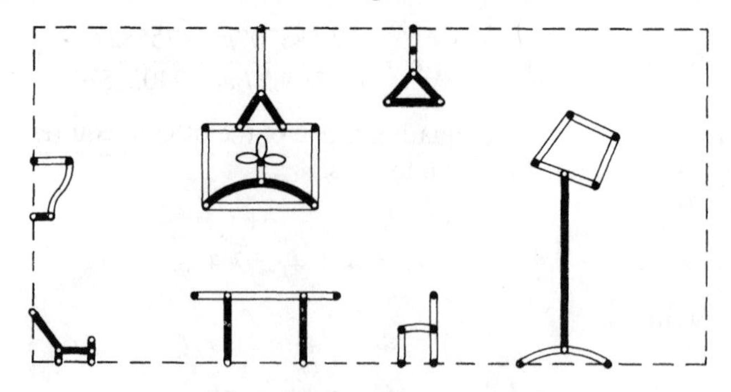

FIG. 9. A restrained Hackenbush room.

At any time when it is his turn to move, Left (Black) may *chop* through any black edge, when that edge disappears, together with any nodes and edges no longer connected to the ground. Right (White) moves in a similar way, by chopping white edges. The game ends when no edge remains to be chopped, and the player unable to move is the loser.

Thus in Fig. 9 Left might start by chopping one leg of the table, which leaves the rest of the table unaffected, but if at his next move he chops the remaining leg, the table disappears. He might alternatively chop away one petal of the flower in the picture—each of these petals is an edge whose two ends coincide. Right's first move might be to chop one of the two white edges supporting the ceiling lamp—of these the lower is the better move, since it leaves him with a further free move. Alternatively, he may chop any edge of the standard lamp except the central column, and so on.

PRELIMINARY DISCUSSION

The positions

0   1 -1 2   2   2 -2 3

have the indicated values. More generally, a position with just n black edges and no white ones will have value n, for Left can take the black edges in a suitable order so as to have n successive moves.

The position 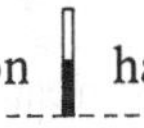 has value $\frac{1}{2}$, for we have the equation

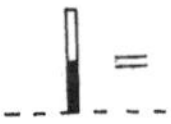

$$= \{\text{---} \mid \ \} = \{0 \mid 1\} = \tfrac{1}{2}$$

and similarly we find the equations

$$= \{\text{---} \mid \ ,\ \} = \{0 \mid 1, \tfrac{1}{2}\} = \tfrac{1}{4}, \quad = \{\text{---}, \ \mid \ \} = \{0, \tfrac{1}{2} \mid 1\} = \tfrac{3}{4}.$$

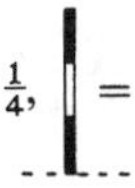

It appears that black edges favour Left, but less so as they get further from the ground, while white edges favour Right in a similar way.

It is not hard to give an inductive proof of the following two propositions. (They must be proved together.)

(i) On chopping a black edge, the value strictly decreases—on chopping a white one it strictly increases.

(ii) The value of every position is a number.

On the other hand, we know no simple rule which enables us to compute this number for an *arbitrary* graph without to some extent playing the game. However, there is a complete theory for *trees*. It turns out that if (P) is some position P, then the *value* of the position (P) depends only on the *value* of P. If the value of P is a *real* number x, then the value of (P) turns out to be the number $1:x$ defined by the conditions:

For real x, the number $1:x$ (the *ordinal sum* of 1 and x) has the first value from the series

$$\frac{x+1}{1}, \frac{x+2}{2}, \frac{x+3}{4}, \frac{x+4}{8}, \frac{x+5}{16}, \ldots$$

for which the numerator of the given expression exceeds 1. (We mean the numerator $x + n$ as written, not the numerator of the number $(x + n)/2^{n-1}$ when written as a rational fraction in least terms.)

In a similar way, the number $(-1):x$ (always negative) will have the first value from the series

$$\frac{x-1,}{1}, \frac{x-2}{2}, \frac{x-3}{4}, \frac{x-4}{8}, \frac{x-5}{16}, \ldots$$

in which the numerator is exceeded by -1, This is the value of the position (P), when P has value x.

Taken together with the obvious result that the value of a position like (P)(Q) is $x + y$, when P has value x and Q value y, these results enable us to evaluate all *trees* in Hackenbush restrained. It is customary to write the values against the edges, in the following way:

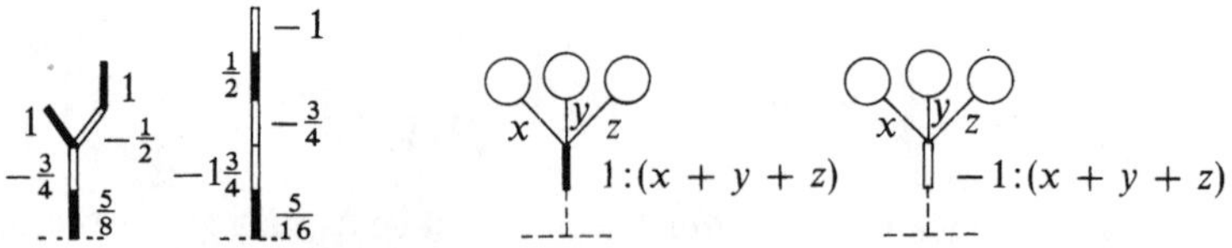

1 1 $-\frac{1}{2}$ $-\frac{3}{4}$ $\frac{5}{8}$ -1 $\frac{1}{2}$ $-\frac{3}{4}$ $-1\frac{3}{4}$ $\frac{5}{16}$ x y z $1:(x + y + z)$ x y z $-1:(x + y + z)$

We explain the occurrence of the functions $1:x$ and $-1:x$ as follows. The moves from the position

(P) are to ------ and (P^L) for Left, (P^R) for Right.

So inductively, the appropriate function is the function $1:x$ defined by

$1{:}x = \{0, 1{:}x^L \mid 1{:}x^R\}$. Now this is a function which maps all numbers onto positive numbers, in order of simplicity. Thus 0, the simplest number, maps to 1, the simplest positive number. Then -1 and 1 map to the simplest positive numbers to the left and right of 1, namely $\frac{1}{2}$ and 2 respectively, and so on. We find under this map that the integers have images as follows

$$\begin{array}{rccccccccccc} x = & -5 & -4 & -3 & -2 & -1 & 0 & 1 & 2 & 3 & 4 & \ldots \\ 1{:}x = & \frac{1}{32} & \frac{1}{16} & \frac{1}{8} & \frac{1}{4} & \frac{1}{2} & 1 & 2 & 3 & 4 & 5 & \ldots \end{array}$$

and then that other real numbers fill in linearly, which explains the above rule.

Note that the rule does not work for *all* numbers. For instance $1{:}(-1/\omega) = 1 - (1/\omega)$ (not $1 - 1/(2\omega)$), and $1{:}\omega = \omega$ (not $\omega + 1$). But the definition in terms of simplicity works for all numbers x, and the inductive definition $1{:}G = \{0, 1{:}G^L | 1{:}G^R\}$ works for all games G.

We postpone further discussion of the properties of this function until Chapter 15, which is its proper home.

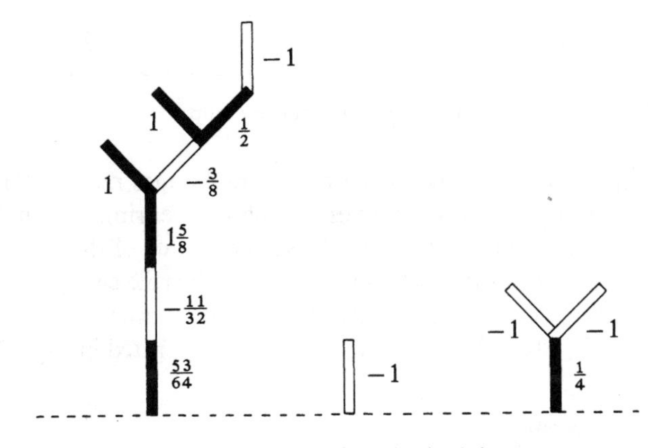

FIG. 10. A restrained Hackenbush forest.

The reader should now be able to see who wins in the position of Fig. 10. Plainly Black—he is exactly five sixtyfourths of a move ahead! (It never ceases to amaze and amuse me that such statements have a precise meaning!)

CHAINS, LOOPS AND INFINITE BEANSTALKS

It follows from the rules for trees that the sign-expansion (Chapter 3) of a *chain* can be read directly from the picture, reading + for black edges, − for white ones, from the ground upwards. So the values of the four chains in Fig. 11 have the sign-expansions

$$+++(+-)+ \;=\; 3{\cdot}11 \;=\; 3\tfrac{3}{4}, \qquad -(-+)- \;=\; -1{\cdot}11 \;=\; -1\tfrac{3}{4}$$

$$+++(+-)++ \;=\; 3{\cdot}111 \;=\; 3\tfrac{7}{8}, \qquad -(-+)-+ \;=\; -1{\cdot}101 \;=\; -1\tfrac{5}{8}$$

where we have bracketed the first sign-change to help the reader apply Berlekamp's rule. Recall that to obtain the binary expansion of the fractional part, for positive numbers we read 0 for $-$, 1 for $+$, and the converse for negative numbers, in either case adding a final 1.

Berlekamp has given a similar rule for the value of a circuit joining the ground to itself (Fig. 11). We *break* the circuit at the node or mid-point of

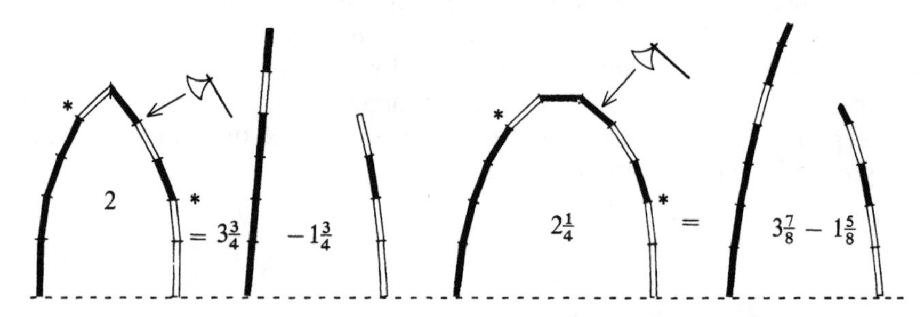

FIG. 11. Berlekamp's rule for loops.

an edge which is midway between the two sign-changes nearest the ground on each side (*s in the diagram), halves of edges appearing (as whole edges) on *both* sides of the fracture when they arise. The value of the circuit is then the sum of the values of its two component parts. The rule can also be applied to a single circuit at some distance from the ground—thus since the value of the left circuit in Fig. 11 is 2, we have the equality illustrated in Fig. 12. But we

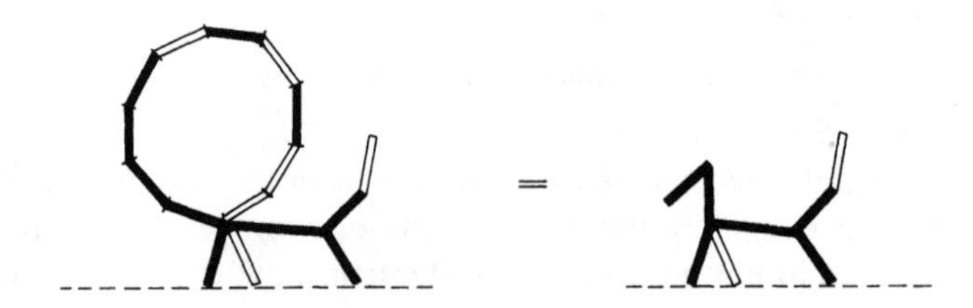

FIG. 12. A head-shrinking equality.

have no general rule for computing values of arbitrary graphs in Hackenbush restrained. Some more information will be given in Chapter 15.

It is perfectly possible to play Hackenbush on infinite trees and certain other infinite graphs, the rules extending naturally. When we do this, arbitrary numbers can arise as values. So for instance the various beanstalks of Fig. 13 have the indicated values.

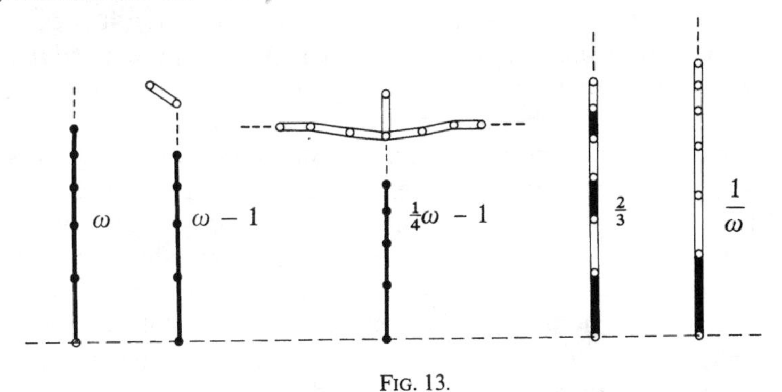

FIG. 13.

THE GAMES OF COL AND SNORT

COL is a map-colouring game introduced by Colin Vout. It is played with a map drawn on a piece of brown paper, a pot of black paint, and a pot of white paint. The players alternately colour countries of the map, subject to the conditions that no country may be coloured twice, and no two countries with a common frontier may be coloured the same colour. Of course, Left uses only black paint, and Right only white.

SNORT is a game introduced by Simon Norton. It is played between two farmers who jointly rent a certain farm, divided into fields. Mr Black buys (black) Bulls, and Mr White (white) coWs, on alternate market days. The animals bought on any one day are to be placed in a field which was previously empty, subject to the condition that no field containing cows may be adjacent to one containing bulls.

If we colour a field black or white according as it contains bulls or cows, we see that both games are played on a map (in the same sense as in the famous 4-colour map problem), the restriction in COL being that adjacent regions may not be similarly coloured, while in SNORT they may not be dissimilarly coloured. This makes it natural to discuss them in similar terms, although as we shall see later, their theories are entirely different.

It is tedious to have to draw complicated maps to specify positions, so we shall simplify the presentation as follows. We discuss COL first. The only effect of a country which has already been painted *black* in COL is to *tint* the neighbouring countries *white*, for these regions may only be painted white in future. Similarly, a *white* painted country causes its neighbours to be tinted *black*. A country that acquires tints of both colours black and white in this way might just as well be erased from the map, since neither player will be allowed to paint it in the future.

In SNORT, these conventions are reversed—any field already coloured causes its neighbours to acquire tints of the *same* colour. But it is still true that a region tinted in both ways can be ignored. Once we have tinted regions according to these conventions, we can ignore all the regions that have actually been *painted*, for they have no further effect on the game.

So we shall represent positions in either of these games by *graphs*, as follows. The graph representing a given position will have a node for each region of that position *which has not already been coloured*, and two nodes corresponding to adjacent regions will be joined by an edge in the graph. The nodes are tinted *black* (●) or *white* (○) or *both* (⊗) or neither (·), and if we like we can omit nodes tinted both black and white. (But the ⊗ notation is still handy.) In Fig. 14 we show the graphs derived in this way from a certain partly coloured map in both COL and SNORT.

There are some further simplifications we can make. An edge joining two oppositely tinted nodes in COL may be omitted, for it has no force (the only effect of any edge is to prevent the nodes at its ends from being similarly coloured). For similar reasons edges joining similarly tinted nodes in SNORT may be deleted. We have also indicated these simplifications in Fig. 14.

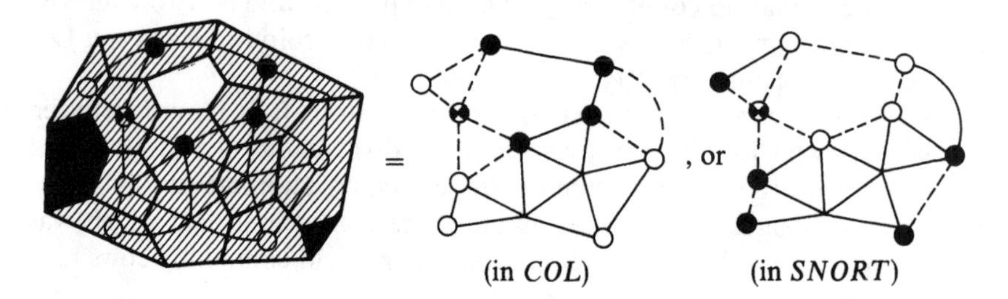

FIG. 14. How maps give graphs.

Simple graphs are now analysed in a manner which should by now be familiar. In the last pages of this chapter we give "dictionaries" for these two games. As well as evaluating simple positions, these dictionaries contain certain general statements which often enable us to simplify very complicated positions not themselves in the dictionary. The methods by which these results are proved will only appear later.

[We might remark at this point that we have found this sort of approach very useful in analysing games in general. One first analyses simple positions, building some kind of dictionary, often in a very unsystematic way. When patterns emerge, if ever, one can often prove general theorems, and then these theorems enable us to 'condense' the dictionary, and on some fortunate occasions, to give a complete theory. Almost all the games used as examples here were first discussed in this way.]

It appears that in COL the values that arise are very restricted in kind. Richard Guy and I have shown that they are all of the form x or $x + *$ for various numbers x. For the inequalities below imply trivially that

$$G^L + * \leqslant G \leqslant G^R + *$$

for any COL position G, and from this the desired result follows by induction. We do not know if denominators of 16 or more can appear in x.

All the values in the COL table can be found by the following sort of analysis. We have the equation

$$= \{0, -2, \tfrac{1}{2} | 3, 1 + *\} = 1.$$

(found by examining the effects of the possible moves), which determines the value of the game on the left hand side in terms of simpler cases.

It is convenient to remember that the simplest number rule in its general form reads:

If there is some number x with $G^L \lhd\!| \, x \lhd\!| \, G^R$ for all G^L, G^R, then G is equal to the simplest such x.

It is also convenient to note the equality $\{x|x\} = x + *$ for all numbers x, which follows from a far more general identity later, and to note that $x + *$ is *greater* than all numbers less than x, *less* than all numbers greater than x, but *incomparable* with x. This also will be generalised later.

Since SNORT values are usually *not* numbers, the SNORT dictionary requires techniques which will be explained later. The abbreviations will also be generalised in Chapters 10 and 15.

A DICTIONARY OF FACTS ABOUT COL

(In general each statement given here has a dual statement in which black and white are interchanged and the inequalities are reversed.)

(1) *Inequalities:* the value of a position is unaltered or increased by *either* tinting a node black (mnemonic: hindering one's opponent is no harm) *or* deleting any edge one end of which has a black tint (mnemonic: let my people go).

(2) *Equalities:* there are many circumstances in which we can say that replacing one configuration by another does not affect the value.

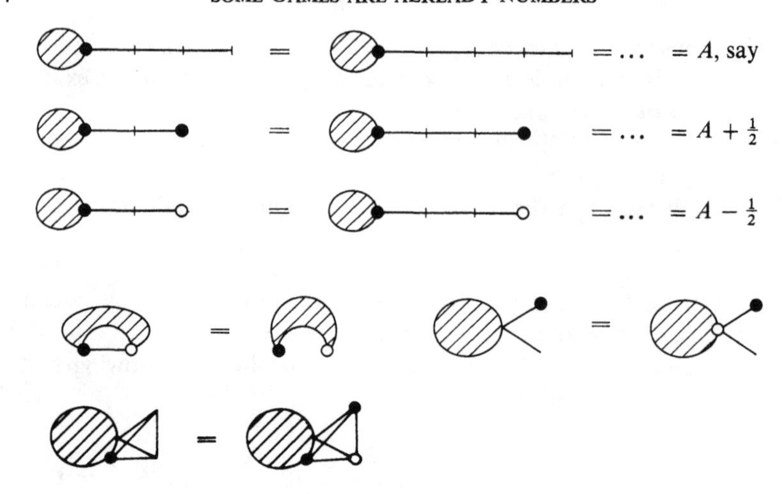

(In general, if two untinted nodes are joined to each other, and to the same set of the remaining nodes, we may tint one black and the other white.)

In general if in some configuration the value is unaltered both when we tint a certain node black and when we tint it white, then that node is "explosive" and may be deleted even when used to join the given configuration to another. So the above equalities are consequences of the following:

Other explosive nodes are indicated by the lightning bolts:

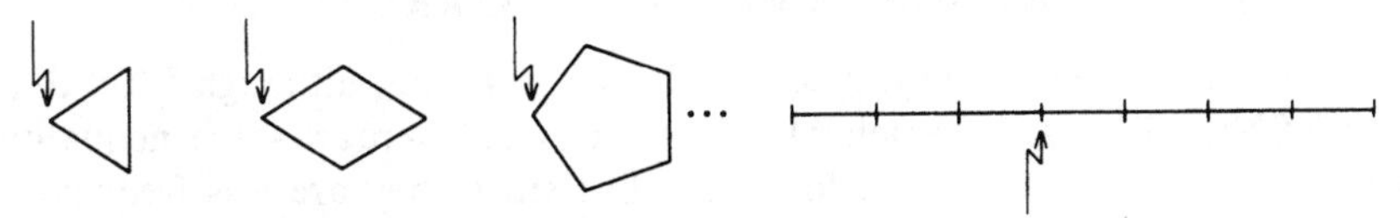

(Any node in an untinted chain with at least three others on each side.)

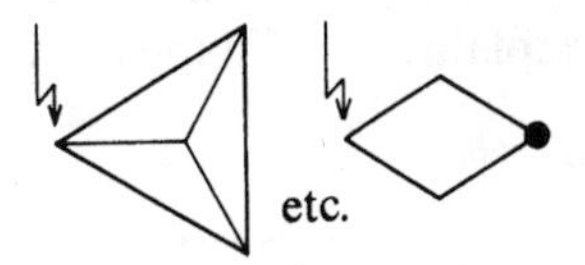

(In each case the explosive node may be tinted without affecting its explosive character.)

Now we list the values of some simple positions (many others can be deduced from these using the above principles and identities):

● = ●—● = ●——● = ... (any length) = 1   + = *

●—+ = ●——+ = ... (any length) = $\frac{1}{2}$ but for longer lengths we have

●—○ = ●——○ = ... (any length) = 0   +—+ = +——+ = ... = 0.

= = $=\frac{1}{4}$

$= \frac{1}{2}$

From these we can deduce the value of any tree with just one tinted node from which lead only a number of chains of untinted nodes.

We can also deduce the corresponding values if the extreme nodes are tinted. (If such a tree is completely untinted, then *either* its central node explodes by one of the above rules, *or* the value is zero.)

$= \frac{1}{2}$ $= 1$ $= 1$ $= \frac{1}{2}$

= $= 0$

(In general a diagram which has a symmetry moving every node and reversing any tints will always have value 0.)

A SHORT SNORT DICTIONARY

It is much harder to do justice to SNORT positions, although I feel that in fact SNORT has a much richer theory than COL. There are some inequality and equality rules like those for COL, but since they are less frequently applicable we do not give many. Perhaps the most valuable rule is that if you can move in a node that is adjacent to every node not your own colour, you should do so. Our abbreviated notation is explained in Chapters 10 and 15.

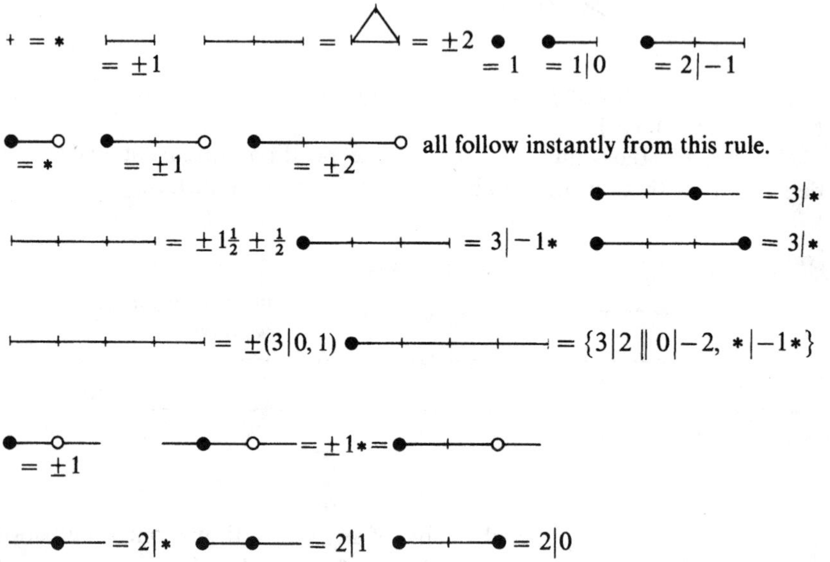

$= *$ $= \pm 1$ $= = \pm 2$ $= 1$ $= 1|0$ $= 2|-1$

$= *$ $= \pm 1$ $= \pm 2$ all follow instantly from this rule.

$= 3|*$

$= \pm 1\frac{1}{2} \pm \frac{1}{2}$ $= 3|-1*$ $= 3|*$

$= \pm(3|0, 1)$ $= \{3|2 \| 0|-2, *|-1*\}$

$= \pm 1$ $= \pm 1* =$

$= 2|*$ $= 2|1$ $= 2|0$

Perhaps it is fortunate that positions in SNORT games tend to break up rapidly, and that we can delete edges joining two nodes with the same tint, so that in practice we need only tabulate the values of small positions. Highly connected positions succumb easily to the above rule, so that in fact it is long chains that are hardest to analyse. The reader should have little difficulty in finding the best move in actual play, even for quite large positions.

Larger COL and SNORT dictionaries will be found in *Winning Ways*.

CHAPTER 9

On Games and Numbers

And now there came both mist and snow,
And it grew wondrous cold:
And ice, mast-high, came floating by,
As green as emerald.

Samuel Taylor Coleridge,
The Ancient Mariner

We know that not all games are numbers, and that for example the game $* = \{0 \mid 0\}$ is not a number, since it is confused with 0. But since for every positive number x, we have $-x < * < x$, and since we have the equality $* + * = 0$, we can confidently handle all games whose values can be expressed as sums of numbers and $*$.

But the position ⊞ in dominoes, which is equivalent to the position +—+ in SNORT, has the rather worse value $\{1 \mid -1\}$. This game G is strictly less than all numbers greater than 1, strictly greater than all numbers less than -1, and confused with all numbers between -1 and 1 inclusive. But fortunately once again, we have $G + G = 0$, so that at least the situation does not get more complicated when we consider multiples of G.

Now in general we can get a lot of information about an arbitrary game G by comparing it with all numbers. The game G will define two "Dedekind sections" in the Class of all numbers (the Left and Right values), and any number between these two sections will be confused with G, while numbers above the greatest or below the least will be comparable with G in the appropriate sense.

This information tells us between which limits G lies, but there is also a *mean value* of G, which tells us where its centre of mass lies. We shall give algorithms for computing the Left, Right, and mean values in this Chapter.

Unfortunately, there is a large part of the argument that is inapplicable to the general infinite game. We adopt the convention of considering only *short* games in detail from now on, until Chapter 16, when we consider the differences between short games and long ones. A *short* game is one which has only finitely many positions in all. But we always explicitly add this adjective to the hypotheses of any theorem which needs it, and often add comments on general games later.

THEOREM 55. (The Archimedean principle.) *For any short game G, there is some integer n with $-n < G < n$.*

[For general G, there is some ordinal α with $-\alpha < G < \alpha$.]

Proof. Take n greater than the total number of positions in G, and consider playing in $G + n$. Left can win this by just decreasing n by 1 each time he moves, waiting for Right to run himself down in G. Since $G + n > 0$, we have $G > -n$, and similarly $G < n$.

[In general we give an inductive proof, taking for α the least ordinal greater than all α_L, α_R.]

THE LEFT AND RIGHT VALUES

We need to know which numbers x have $x \geqslant G$, and which y have $y \leqslant G$. These conditions define two *Dedekind sections* in the Class of all numbers, called the *Left section* $L(G)$ and the *Right section* $R(G)$, as follows.

A number x is put into the right-hand part of $L(G)$ iff $x \geqslant G$, and so in the left-hand part if $x \lhd\!| \, G$, while y is put into the left part of $R(G)$ if $y \leqslant G$, the right part if $y \, |\!\rhd G$.

In particular, if z is any number, $L(z)$ has for its left part all numbers strictly less than z, z and greater numbers forming its right part, while $R(z)$ has z and smaller numbers to its left, greater numbers to its right.

So $L(z)$ and $R(z)$ are the sections just to the left and right of z, respectively. For a more general game G, if $L(G)$ is one of the two sections $L(x)$, $R(x)$ for some number x, we call x the *Left value* $L_0(G)$ of G, while y is called the *Right value* $R_0(G)$ if $R(G) = L(y)$ or $R(y)$.

We introduce the obvious order on sections ($S < T$ if some number is to the right of S and the left of T), so that $L(z) < R(z)$ for each number z. But for other games, the inequality goes the other way, for if $L(G) < x < R(G)$, we have $x \leqslant G \leqslant x$, and so $G = x$. How do we compute these sections, in general?

THEOREM 56. *We have* $L(G) = \max_{G^L} R(G^L) = L$, *say,*

$$\text{and } R(G) = \min_{G^R} L(G^R) = R, \text{ say}$$

unless $L < R$, when G is a number, namely the simplest number x satisfying $L < x < R$, when we have $L(G) = L(x)$, $R(G) = R(x)$.

[For general G, we must replace max and min by sup and inf.]

Proof. We tackle the case $L < R$ first. If x is the simplest number between, then

$$x^L < L < x < R < x^R,$$

so the moves from $G - x$ to $G - x^L$, $G - x^R$ are no good. But neither, in view of the definition of L and R, are those to $G^L - x$ and $G^R - x$, so that $G - x$, having no good move for either player, is a zero game.

In the case that $L \geqslant R$, the moves to $G^L - x$, $G^R - x$ are bad for the same reason, if $x > L$, $x < R$, respectively. So we need only consider, if $x > L$, moves to $G - x^R$, and if $x < R$, moves to $G - x^L$. But these fail, since we have $x^R > x > L$ in the first case, and $x^L < x < R$ in the second.

STOPPING POSITIONS

When the value of a position *is* a number, neither player will wish to move in it, for any move by Left will decrease the value, and any move by Right increase it. We can be kind to the players and agree to *stop* the game (possibly before its real end) as soon as the value becomes a number, and score positive values in favour of Left, negative ones in favour of Right. So we shall call positions of G which are equivalent to numbers the *stopping positions* of G.

Now Left will naturally prefer to arrange that when the game stops in this sense, its value will be as large as possible, while Right will prefer to make it small. If they play in this way, the value of the game when its stops will be a perfectly definite number which depends only on who starts. Moreover, each player will prefer that when the game stops it is his opponent who is about to move (and so do himself some harm).

Now we can describe the situation by saying that if Left starts, the game will end at some number x, with some player P (Left or Right) about to play, by the equality $L(G) = P(x)$, and the corresponding assertion that if Right starts the game will end at a number y with Q about to play, by the equality $R(G) = Q(y)$. This is because Theorem 56 tells us that the Left and Right sections of G are computed exactly as we should compute the numbers x and y, and locate the players P and Q.

Summary. We can determine exactly what are the order relations between a game G and *all* numbers by simply playing G intelligently until it stops and then noting the value and who is about to play.

Examples

The game $\{5 \mid 4, 7\}$. In this game, if Left starts, the game will end at 5, with Right to play, and so $L(G) = R(5)$, the section "just to the right" of 5.

If Right starts, the game ends with Left to play, at the number 4, if Right has any sense, and so $R(G) = L(4)$, just to the left of 4. We conclude that G is *strictly less* than all numbers greater than 5, *strictly greater* than all numbers less than 4, and *confused* with all numbers between 4 and 5 inclusive.

The game $\{9 \mid \{7 \mid 2\}\}$. Here $L(G) = R(9)$, the argument being as before, but we have $R(G) = R(7)$, for if Right starts, moving to $\{7 \mid 2\}$, Left continues the game for one more move, before it stops at value 7 with Right to play. So the game is *less* than numbers greater than 9, *greater* than numbers less than *or equal to* 7, and *confused* with numbers between 7 (exclusive) and 9 (inclusive).

The game $\{\{3 \mid 0\} \mid \{\frac{1}{2} \mid 9\}\}$. Here if Left starts we arrive at $L(0)$, while if Right starts we stop at $R(\frac{1}{2})$. But these are *not* the Left and Right sections of G, for we have $R(\frac{1}{2}) > L(0)$. So in this case, G is a number, namely the simplest number x satisfying $L(0) < x < R(\frac{1}{2})$, namely 0 itself. So in fact we have $L(G) = L(0)$, $R(G) = R(0)$, $G = 0$.

If we had replaced the position 0 here by $\frac{1}{4}$, the answer would have been $\frac{1}{2}$; if by -1, the answer would still have been 0; and if by $+1$, we would no longer have had a number, and $L(G) = L(1)$, $R(G) = R(\frac{1}{2})$.

Moral. When computing Left and Right values, look out for the inequality $L < R$ between Left and Right sections.

The games $*$ *and* $\uparrow$. Since $* = \{0 \mid 0\}$, we have $L(*) = R(0)$, $R(*) = L(0)$. We need not beware, since L is safely greater than R, and we conclude that $*$ is greater than all negative numbers, less than all positive numbers, but confused with 0. Again, since $\uparrow = \{0 \mid \{0 \mid 0\}\}$, we find $L(\uparrow) = R(0)$, $R(\uparrow) = R(0)$, and so $\uparrow$ is strictly positive (as we knew) but strictly less than all positive numbers. (Note that for $\uparrow$, we had $L = R$, so *almost* had to beware, etc. But not quite!)

So these games are infinitesimal in a totally new sense, for we have, for instance,

$$0 < \uparrow < \frac{1}{\omega}, \quad 0 < \uparrow < \frac{1}{\varepsilon_0}, \quad 0 < \uparrow < \frac{1}{2^{\aleph_0}}, \ldots$$

($2^{\aleph_0}$ being identified with the smallest ordinal having that cardinal), and so on. (Informally, $0 < \uparrow < 1/\mathbf{On}$.) Rather than invent some long adjective to qualify the word infinitesimal in this sense, we simply call such games *small*. So a *small* game is any game G for which we have $-x < G < x$ for every possible positive *number* x. Some small games (like $\uparrow$) are positive, others (like $\downarrow$) negative, and still others (like $*$) are fuzzy, while of course zero is itself a small game. So the small World is indeed a microcosm of the larger one.

THE ALL SMALL GAMES

We call a game *all small* if all its positions are small games.

THEOREM 57. *G is all small if and only if every stopping position of G is zero.*

Proof. If some position of G were a non-zero number, it would be a non-small position of G. So we need only prove that if all the stopping positions are zero, then so are the Left and Right values. This follows immediately from Theorem 56.

Note. There are positive games smaller than all positive all small games. One such is the value $\{0 \mid \{0 \mid -2\}\}$ of the domino position 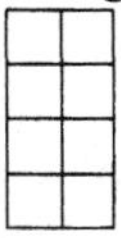. The multiples of $\uparrow$ are among the largest of all small games.

THE MEAN VALUE THEOREM

We shall prove that for every short game G there is a real number m, called the *mean value* $m(G)$, such that for every finite n, the game nG is "nearly equal" to nm. This result, for a slightly different class of games, was first conjectured by J. Milnor, and first proved for that class by O. Hanner. A simplified proof, for the Class of games considered here, was given by Elwyn Berlekamp. All these proofs depend on a fairly complicated analysis that yields a strategy for playing nG so as to ensure a stopping value near the desired mean value nm.

The first proof given here is the remarkable "1-line" proof found by Simon Norton, which proves the existence of the mean value and finds good bounds for nG, but which does not enable us to compute this value! Then we shall give another proof, found by Norton and the author jointly, which gives us an easy algorithm for computing the mean value and much other information. This new proof formalises and simplifies an idea whose germ is found in the papers of Milnor and Hanner but which was discovered only after a completely independent analysis.

We start with some obvious inequalities about the Left and Right values $L_0(G)$, $R_0(G)$. Recall that these are the numbers next to the sections $L(G)$ and $R(G)$.

THEOREM 58. *We have*

$$R_0(G) + R_0(H) \leqslant R_0(G + H) \leqslant R_0(G) + L_0(H) \leqslant L_0(G + H) \leqslant L_0(G) + L_0(H).$$

Proof. These are obvious in terms of strategies. Thus Left, playing second in $G + H$, can guarantee a stopping value of at least $R_0(G) + R_0(H)$ by replying always in the component Right moves in, and following in that component his strategy yielding its Right value. The others can be proved similarly, but are in fact equivalent to this one. For instance

$$R_0(G) = R_0(G + H - H) \geqslant R_0(G + H) + R_0(-H) = R_0(G + H) - L_0(H).$$

THEOREM 59. (*The mean value theorem.*) *For every short game G there is a number m(G) and a number t (both real) such that*

$$nm(G) - t \leqslant nG \leqslant nm(G) + t$$

for all finite integers n.

Proof. After the previous theorem, it will suffice to prove that $L_0(nG)$ and $R_0(nG)$ have a difference bounded independently of the number n, for then $(1/n)R_0(nG)$ and $(1/n)L_0(nG)$ must converge to a common value $m(G)$, since we have the inequalities

$$R_0(G) \leqslant \frac{1}{n} R_0(nG) \leqslant \frac{1}{n} L_0(nG) \leqslant L_0(G).$$

But we have

$$R(nG) \leqslant L(nG) = R((n - 1)G + G^L) \leqslant R(nG) + L(G - G^L)$$

for the G^L for which the max in Theorem 56 is attained.

Note. The proof shows also that the number t is bounded by max $L_0(G - G^L)$, and similarly, bounded by max $L_0(G^R - G)$. These inequalities will be improved later.

THE TEMPERATURE THEORY

We can regard the game G as vibrating between its Right and Left values in such a way that on average its centre of mass is at $m(G)$. So in order to compute $m(G)$ we must find some way of cooling it down so as to quench these vibrations, and perhaps if we cool it sufficiently far, it will cease to vibrate at all, and *freeze* at $m(G)$.

Now the heat in a game comes largely from the excitement of playing it—if there are positions in G from which each player can gain tremendously by making a suitable move, then G will naturally be very heated! So for instance the game $\{1000 \mid -1000\}$ is a very hot position, for although its mean value is zero, the player who moves first in it stands to gain 1000. On the natural scale, the temperature of this game is 1000°.

On this theory, we should be able to cool G through a temperature of $t°$

by making it just that much less exciting to move in each position of G that has not already stopped. So we shall define a new game G_t (*G cooled by t*) by charging each player a fee of t every time he makes a move, until the value becomes a number. A formal definition is complicated slightly by the need to detect when this has taken place.

Definition. If G is a short game, and t a real number $\geqslant 0$, then we define the *cooled game* G_t by the formula

$$G_t = \{G^L{}_t - t \mid G^R{}_t + t\},$$

unless possibly this formula defines a number (which it will for all sufficiently large t). For the smallest values of t for which this happens, the number turns out to be constant (that is, independent of t), and we define G_t to be this constant number for all larger t.

[The reader will see that our definition of G_t contains an assertion, and so does not really count as a definition until this assertion is verified to hold for all short G. The reason the theory does not work for general games G is that this assertion fails to hold for certain long games G.]

To see how the definition works, we treat the case $G = \{4 \mid 1\}$, supposing it already established that $4_t = 4$, $1_t = 1$ for all t. Then our formula gives $G_t = \{4 - t \mid 1 + t\} = G(t)$ unless perhaps when $G(t)$ is a number, when...? When is $G(t)$ a number? Obviously when t exceeds $1\frac{1}{2}$. What number is $G(t)$? The answer to this question depends on t, and in fact we have

$$\begin{aligned} G(t) &= 2\tfrac{1}{2} \text{ for } 1\tfrac{1}{2} < t \leqslant 2 \\ & 2 \text{ for } 2 < t \leqslant 3 \\ & 1 \text{ for } 3 < t \leqslant 4 \\ & 0 \text{ for } 4 < t. \end{aligned}$$

So as the definition asserts, $G(t)$ is a constant number ($2\frac{1}{2}$) for all the smallest numbers t for which it is a number (namely the numbers t with $1\frac{1}{2} < t \leqslant 2$), and so we have $G_t = \{4 - t \mid 1 + t\}$ for $0 \leqslant t \leqslant 1\frac{1}{2}$, and $G_t = 2\frac{1}{2}$ for *all* larger t.

We define the sections $L_t(G)$ and $R_t(G)$ to be $L(G_t)$ and $R(G_t)$.

THEOREM 60. *For all short games G and real numbers $t \geqslant 0$, we have*

$$L_t(G) = \max R_t(G^L) - t = L_t, \textit{ say,}$$

and

$$R_t(G) = \min L_t(G^R) + t = R_t, \textit{ say,}$$

unless possibly $L_t < R_t$. In this latter case, G_t is a number x, namely the simplest *number between L_u and R_u for all small enough u with $L_u < R_u$, and we then have* $L_t(G) = L(x)$, $R_t(G) = R(x)$.

Proof. This follows immediately on applying Theorem 56 to G_t. For the moment, we are continuing to suppose that G_t is well-defined.

THE THERMOGRAPH OF *G*

We find it convenient to describe the various numbers associated with G on a diagram. The Left options of the G with which we are concerned will usually be greater than the Right ones, so we shall reverse the normal convention and put positive values on the left, and negative ones on the right. (This happy convention has various other advantages which will appear gradually.) The temperature scale is vertical, and at height t we indicate the Left and Right values of G_t, which define the Left and Right boundaries of the *thermograph* of G. (We are indebted to Elwyn Berlekamp for this snappy substitute for our own phrase "thermal diagram".)

As our example, we take the game $G = \{\{7 \mid 5\} \mid \{4 \mid 1\}\}$. The calculation of the thermal properties of this game is illustrated in Fig. 15, the game itself being drawn below its thermograph. Since the games 7, 5, 4 and 1 are already numbers, they remain constant when cooled by arbitrary t, so that their thermographs are vertical lines above the appropriate numbers.

Now the Left boundary $L_t(H)$ for the game $H = \{7 \mid 5\}$ is obtained, at any rate until H_t becomes a number, by subtracting t from the Right boundary of the game 7. Since this is vertical, and subtraction corresponds to moving

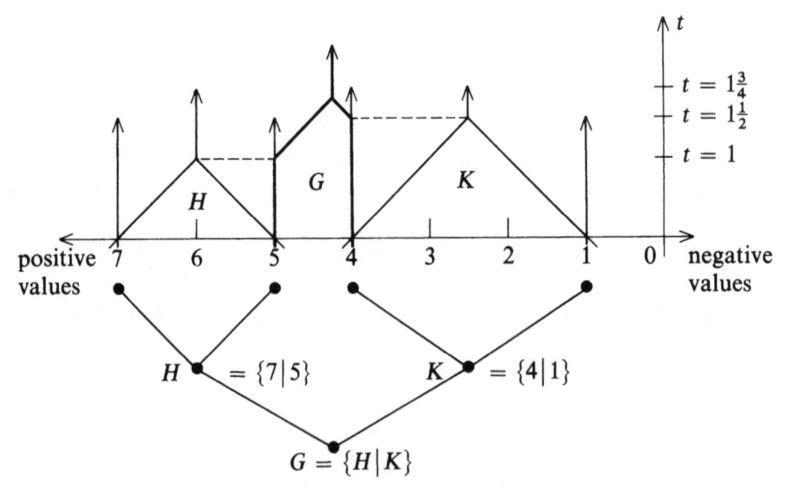

FIG. 15. Computing thermographs.

right in the diagram, this gives a line starting at 7 and moving diagonally up and right. Similarly the Left boundary is a line starting at 5 and initially moving diagonally up and left. But since these lines meet at a height of 1 above the number 6, H_t will be the constant number 6 for all t larger than 1, and the Left and Right boundaries will be vertical above this point.

So the thermograph of H is *the pyramid* $\diagup 7, 5 \diagdown$—that is to say, an isosceles right-angled triangle with hypotenuse on this interval, except that, like all thermal diagrams, it has a mast on top. The Right boundary of this diagram consists of the right side of the triangle together with the mast.

In a similar way, the game $K = \{4 \mid 1\}$ yields the pyramid $\diagup 4, 1 \diagdown$, with a mast which starts at a height of $1\frac{1}{2}$ above the point $2\frac{1}{2}$. Its Left boundary is the left side of this pyramid together with the mast. Now we compute $L_t(G) = R_t(H) - t$, $R_t(G) = L_t(K) + t$ (until G_t becomes a number) by pushing the Right boundary of H still further right, and the Left boundary of K still further left. Applied to the Right boundary of H this yields a line starting at 5 and travelling vertically upwards until $t = 1$, then diagonally right and up thereafter. From the Left boundary of K we get a line vertical till $t = 1\frac{1}{2}$, then diagonally up and left.

These lines meet at a height $t = 1\frac{3}{4}$ directly above the value $4\frac{1}{4}$, and so they define the Left and Right boundaries of G below this point, these boundaries above this point being vertical. So the diagram for G is a lop-sided "house" with a mast.

When we consider the implications of this procedure for the general short game G, we obtain:

THEOREM 61. *For any short game G, the thermograph is a region whose Left boundary is a line proceeding either vertically or diagonally up and right in stretches, the Right boundary being in stretches vertical or diagonal up and left. Beyond some point, both boundaries coincide in a single vertical line (the mast). The coordinates of all corners in the diagram are dyadic rationals.*

Proof. This requires only the observation that on subtracting t from a line which is vertical or diagonal up-and-left we obtain one correspondingly diagonal up-and-right or vertical, and that two such lines aiming towards each other must meet, at a point whose coordinates can be found with a single division by 2.

The proof of the theorem assures us at last that the definition of G_t has the properties presupposed in it, and incidentally makes Theorem 60 an honest theorem.

Now we ask about the corresponding sections $L(G_t)$ and $R(G_t)$. On which side are they of the numbers near to them?

THEOREM 62. (See Fig. 16). *The sections $L(G_t)$ and $R(G_t)$ are "just inside"* *the boundary of the diagram on vertical stretches, "just outside" on diagonal stretches. At points of the mast above its foot, $L(G_t)$ is to the right of $R(G_t)$ in the diagram; that is to say, $L(G_t) < R(G_t)$. At corners of the diagram the sections behave in the same way as at immediately smaller values of t. (So their behaviour is "continuous downwards".)*

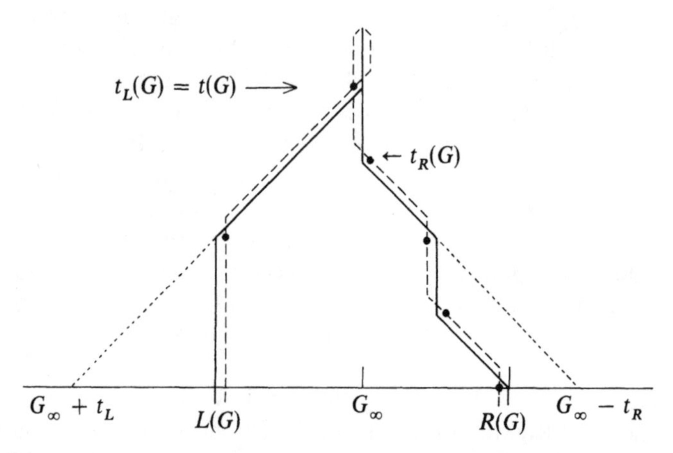

FIG. 16. The left and right sections of G_t are indicated by the dashed lines. Note how they cross the firm lines at corners, and cross each other at the foot of the mast. This behaviour is typical.

Proof. These properties are preserved in the passage from the diagrams for G^L and G^R to that for G.

Now Theorem 62 makes it natural to prolong the boundaries just a little way downward below the line $t = 0$. These prolongations are to be vertical when the corresponding section at $t = 0$ is just inside the thermograph diagram, and diagonally "outwards" when it is just outside. When we do this (as we shall), we read off the nature of the sections for $t = 0$ from the diagram as well. The rules for computing these prolongations are the obvious extensions of the rules for the rest of the diagram, and we shall say no more about them. The reader who examines Figs 15 and 16 closely will see that these prolongations were already present.

THEOREM 63. $\quad G \geqslant x$ *implies* $G_t \geqslant x$

$$(x + G)_t = x + G_t$$

$$(x - G)_t = x - G_t$$

for all short games G and dyadic rationals x.

Proof. Obvious from the properties and construction of thermographs.

THEOREM 64. *$(G + H)_t = G_t + H_t$ for short G, H.*

Proof. If G, H, or $G + H$ is equal to a number x, this follows from Theorem 63. Otherwise, we can use the inductive definitions of G_t, H_t, $(G + H)_t$ to give a 1-line proof:

$$\begin{aligned} G_t + H_t &= \{G^L{}_t - t + H_t, G_t + H^L_t - t \mid G^R{}_t + t + H_t, G_t + H^R{}_t + t\} \\ &= \{(G + H)^L_t \mid (G + H)^R_t\} = (G + H)_t. \end{aligned}$$

THEOREM 65. *If $G \geqslant H$, then $G_t \geqslant H_t$. In particular, from $G = H$, we can deduce $G_t = H_t$.*

Proof. We have $G \geqslant H$ iff $G - H \geqslant 0$, so this theorem follows from the previous one.

Note. The contrary possibility that the value of G_t might depend on the *form* of G makes Theorems 63 and 64 slightly more subtle than they appeared at first sight. But all is now well.

Definition. We write G_∞ for the ultimate value of G_t, and t_L for the value of t beyond which $L(G_t) = L(G_\infty)$, t_R for the value beyond which $R(G_t) = R(G_\infty)$. The numbers t_L and t_R are called respectively the *Left and Right temperatures* of G, and their maximum is just *the* temperature $t(G)$ of G. See Fig. 16.

THEOREM 66. *G_∞ is none other than the mean value $m(G)$ of G. (From now on, we use the new notation G_∞.) We have the inequalities*

$$L(G_t) \leqslant L(G) \leqslant L(G_t) + t$$

$$R(G_t) - t \leqslant R(G) \leqslant R(G_t)$$

$$t(G + H) \leqslant \max\,(t(G), t(H))$$

(and similar inequalities with $t(G)$ replaced by $t_L(G)$, $t_R(G)$), and also the equalities

$$t_L(G) = t_R(-G), t(G) = t(-G),$$

and the "cooling equality"

$$(G_t)_u = G_{t+u}.$$

Proof. The first statement follows from Theorem 64 and the facts that $L(G_t) \leqslant L(G)$, $R(G) \leqslant R(G_t)$, which, like the remaining inequalities of the next two lines follow from the assertions about the slopes of the Left and Right boundaries. The third inequality is proved as follows: since for $t > t(G), t(H)$ we have $G_t = G_\infty$, $H_t = H_\infty$, for such t we have

$$(G + H)_t = G_t + H_t,$$

a number. So such t are also greater than $t(G + H)$. The inequalities about $-G$ are obvious. So we are left with the cooling equality, which has a 1-line inductive proof.

This theorem implies in particular that we obtain the thermograph for G_t by *submerging* that for G to the depth t (see Fig. 17). In other words, the way we cool a game is by pouring cold water on it!

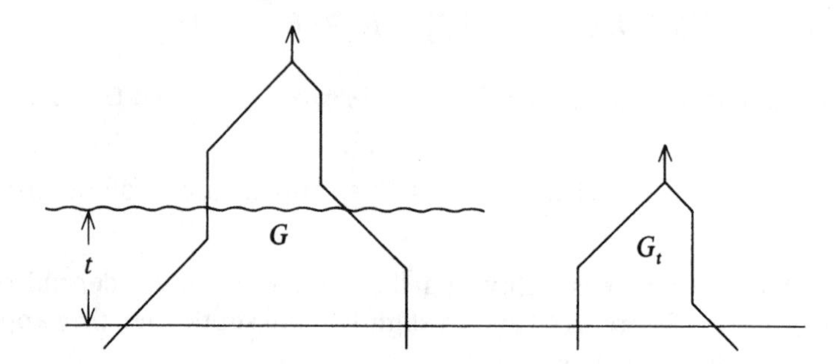

FIG. 17. How to cool a game by pouring water on it.

"Thermography" has been much extended and generalised by Elwyn Berlekamp and his co-workers, who have applied it to "Go" and other traditional games in the following works:

E. R. Berlekamp, Blockbusting and Domineering, *J. Combin. Theory Ser. A*, **49**(1988) 67–116.

Elwyn Berlekamp, Introduction to Blockbusting and Domineering, in *The Lighter Side of Mathematics* (R. K.Guy & R. W. Woodrow, eds.), Spectrum Series, Math. Assoc. of America, 1994, 137–148.

Elwyn Berlekamp, An economist's view of combinatorial games, in *Games of No Chance*, Proc. MSRI Workshop on Combinatorial Games, July 1994, Berkeley CA (Richard Nowakowski, ed.), MSRI Publ. **29**, Cambridge University Press, pp. 365–405.

Elwyn Berlekamp & David Wolfe, *Mathematical Go: Chilling Gets the Last Point*, A K Peters, Ltd., 1994; also as *Mathematical Go: Nightmares for the Professional Go Player*, Ishi Press International, 1994.

Dan Calistrate, The reduced canonical form of a game, in R. J. Nowakowski (ed.), *Games of No Chance*, Proc. MSRI Workshop on Combinatorial Games, July 1994, Berkeley CA (Richard Nowakowski, ed.), MSRI Publ. **29**, Cambridge University Press, pp.409–416.

Olof Hanner, Mean play of sums of positional games, *Pacific J. Math*, **9**(1959) 81—99; *MR* **21** #3277.

Kao Kuo-Yuen, Sums of hot and tepid combinatorial games, PhD thesis, Univ. of North Carolina, Charlotte, 1997, 115 pp.

David Moews, On some combinatorial games connected with Go, PhD thesis, Univ. of California, Berkeley, Dec. 1993.

Martin Müller, Elwyn R. Berlekamp & William L. Spight, Generalized thermography: algorithms, implementation and application to Go endgames, TR-96-030. Internat. Comput. Sci. Inst., Berkeley CA, ISSN 1075-4946, 1996.

William M. Spight, Extended thermography for multiple kos in Go, in *Computers and Games*, Springer Lecture Notes in Computer Science, **1558**(1999) 232–251.

Laura J. Yedwab, On playing well in a sum of games, MSc thesis, MIT, 1985. Issued as MIT/LCS/TR-348, MIT Lab. for Comput. Sci., Cambridge MA.

CHAPTER 10

Simplifying Games

You boil it in sawdust: you salt it in glue:
You condense it with locusts and tape:
Still keeping one principal object in view—
To preserve its symmetrical shape.

Lewis Carroll, "The Hunting of the Snark"

One quite valuable way to simplify games is to simplify our notation for them! (This is more important than it might seem, because even with the best will in the world, the names of games can get inordinately long.) So we first present some useful abbreviations.

We omit the curly brackets round games whenever this is possible without too much confusion—so for instance we shall write $A, B \mid C$ for the game $\{A, B \mid C\}$. Next, we need some way of distinguishing between $\{\{A \mid B\} \mid C\}$ and $\{A \mid \{B \mid C\}\}$, and so we introduce $\|$ as a 'stronger' separator than $|$, when these games become $A \mid B \| C$ and $A \| B \mid C$ respectively. ($A \| B \mid C$ may be pronounced "A slashes B slash C".) Thus the game we used as an example for temperature theory would now be called $7 \mid 5 \| 4 \mid 1$. Sometimes it is handy to introduce triple slashes $|||$, but usually we can get along quite happily with judicious use of brackets to supplement the above conventions.

The initial positions of many games are of the form

$$\{A, B, C, \ldots \mid -A, -B, -C, \ldots\}$$

being symmetrical as regards Left and Right. So we introduce the abbreviation $\pm(A, B, C, \ldots)$ for this game. In particular, the notation $\pm G$ will mean $\{G \mid -G\}$. Note that this will prevent us in future from using $\pm$ to denote an ambiguous sign, so that the phrase "$+x$ or $-x$" will appear more commonly than usual from now on. Finally, there are many positions of the form

$$\{A, B, C, \ldots \mid A, B, C, \ldots\}$$

in which the moves for Left and Right are *identical*, rather than *symmetrical.*

We shall use

$$\{A, B, C, \ldots\}$$

as an abbreviation for this game.

Some other notational conventions for particular games will be introduced in our Chapter 15. A fairly complete dictionary is given at the end of the book.

However, the real simplifications we have in mind concern the form of G rather than its name. The main problem is to see how we can simplify the *form* of a given game without affecting its *value*. We first discuss some modifications which might change the value, but in a predictable way.

THEOREM 67. *The value of G is unaltered or increased when we*
(*i*) *increase any G^L or G^R,*
(*ii*) *remove some G^R or add a new G^L,*
(*iii*) *replace the G^R by the K^R, for any game $K \geqslant G$.*

Proof. Let H be the game obtained by so modifying G. Then in the game $H - G$ it is easy to check that Right has no good first move.

Informally, it is even more obvious that these modifications are in Left's favour, for giving him new moves or prohibiting certain moves for Right will not harm Left. These principles are used repeatedly in analysing individual games, often in very much more general forms.

DOMINATED AND REVERSIBLE OPTIONS

Suppose two different Left options of G are comparable with each other, say $G^{L_1} \leqslant G^{L_0}$. Then we say G^{L_1} is *dominated* by G^{L_0}, since Left will plainly regard the latter as the better move. Similarly, if $G^{R_1} \geqslant G^{R_0}$ (note the reversed inequality) we call G^{R_1} dominated by G^{R_0}.

Now suppose instead that the Left option G^{L_0} has itself a Right option $G^{L_0R_0}$, say, for which we have the inequality $G^{L_0R_0} \leqslant G$. Then we say that the move from G to G^{L_0} is a *reversible* move, being reversible through $G^{L_0R_0}$. Similarly a Right option G^{R_1} of G is reversible (through $G^{R_1L_1}$) if and only if it has some Left option $G^{R_1L_1} \geqslant G$. It turns out that whenever one player (Left, say) makes a reversible move, his opponent might as well reverse it (for he improves on the original position by so doing). So instead of moving from G to G^{L_0}, Left might as well move straight from G to some $G^{L_0R_0L}$. A formal version of this result is part of the next theorem.

THEOREM 68. *The following changes do not affect the value of G.*
(*i*) *inserting as a new Left option any $A \lhd\!| \; G$, or as a new Right option any $B \; |\!\rhd G$.*

(ii) *Deleting any dominated option*
(iii) *If G^{L_0} is reversible through $G^{L_0R_0}$, replacing G^{L_0} as a Left option of G by all the Left options $G^{L_0R_0L}$ of $G^{L_0R_0}$.*
(iv) *If G^{R_1} is reversible through $G^{R_1L_1}$, similarly replacing G^{R_1} by all the $G^{R_1L_1R}$.*

Proof. Because of the importance of this theorem, we give a more detailed proof. Suppose first that $A \lhd\!\mid G$, and let $H = \{G^L, A \mid G^R\}$ be the modified game in (i). Then in $H - G$ the moves from H to G^L, G^R have as counters those from $-G$ to $-G^L$, $-G^R$, and conversely, and the move from H to A yields the position $A - G \lhd\!\mid 0$ by assumption. So there is no good move in $H - G$, whence $H = G$.

Part (ii) now follows, for if G^{L_1} is dominated by G^{L_0}, and H denotes G with G^{L_1} deleted, we have $G^{L_1} \leqslant G^{L_0} \lhd\!\mid H$, and so the insertion of G^{L_1} will not affect the value of H. Recall the fact that for any game G and any G^L, G^R, we have $G^L \lhd\!\mid G \lhd\!\mid G^R$, for from the difference $G - G^L$ or $G^R - G$, Left can plainly move to 0. (This theorem is part of Theorem 0 of part 0!)

Part (iii) is the most important and least obvious part. Let us write $G = \{G^{L_0}, G^{L'} \mid G^R\}$, $H = \{G^{L_0R_0L}, G^{L'} \mid G^R\}$, where $G^{L'}$ denotes the typical Left option other than G^{L_0} of G. Now consider the difference

$$H - G = \{G^{L_0R_0L}, G^{L'} \mid G^R\} + \{-G^R \mid -G^{L'}, -G^{L_0}\}.$$

The moves from H to $G^{L'}$ or G^R and from $-G$ to $-G^{L'}$, $-G^R$ counter each other, so we need only consider those from H to $G^{L_0R_0L}$ and from $-G$ to $-G^{L_0}$. The first of these is shown to be bad by

$$G^{L_0R_0L} \lhd\!\mid G^{L_0R_0} \leqslant G,$$

and the second is countered by the move from $-G^{L_0}$ to $-G^{L_0R_0}$, after which Right is to move in the position $H - G^{L_0R_0}$. His moves from $-G^{L_0R_0}$ to $-G^{L_0R_0L}$ have counters in H, so he must move from H to G^R. But this is a bad move, since $G^R - G^{L_0R_0} \geqslant G^R - G \mid\!\rhd 0$.

Part (iv) follows by symmetry.

THE SIMPLEST FORM OF A SHORT GAME

Now let G be a short game. We aim to find the simplest form of G. By induction, we can suppose that each game G^L, G^R has already been put into simplest form, if we like. In any case, we proceed as follows—eliminate from G any option which is dominated by some other option, and then replace any reversible option G^{L_0} or G^{R_1} by the corresponding smaller positions $G^{L_0R_0L}$ or $G^{R_1L_1R}$, respectively. Repeat, if necessary, until no option of G is dominated or reversible.

THEOREM 69. *Suppose that G and H (not necessarily short) have neither dominated nor reversible options. Then G and H are equal if and only if each Left or Right option of either is equal to a corresponding option (Left or Right respectively) of the other.*

Proof. Suppose $G = H$, and consider playing $G - H$. The move for Right to $G^R - H$ must have a reply for Left, say to either $G^{RL} - H$ or $G^R - H^R$. The former case is impossible, for it implies $G^{RL} \geqslant H = G$, so that G^R was reversible in G. So we have proved that for each G^R there is some H^R with $G^R \geqslant H^R$. Since similarly each $H^R \geqslant$ some G^R, and neither game has dominated options, we must in fact have each $G^R =$ some H^R and conversely. Similar statements hold for the Left options.

This theorem assures us that each short game has a unique *simplest form.* We shall now discuss some examples.

Examples. The position $\{\uparrow \mid \uparrow\}$. We know already that

$$\{\uparrow \mid \uparrow\} \geqslant \{0 \mid \uparrow\} = \uparrow + \uparrow + *$$

obviously greater than $* = \uparrow^R$. So $\uparrow$ is reversible through $*$ as a Left option, and can therefore be replaced by $*^L = 0$. So we have $\{\uparrow \mid \uparrow\} = \{0 \mid \uparrow\}$. Since there is no 0^R, 0 cannot be reversible in this (indeed, 0 can never be reversible in any game), and since $\{0 \mid \uparrow\}$ is positive (Left can win, Right can't), $\uparrow$ is not reversible as a Right option. So $\{\uparrow \mid \uparrow\} = \{0 \mid \uparrow\}$ in simplest form.
(Recall that $\uparrow$ is the game $\{0 \mid *\}$, where $* = \{0 \mid 0\}$.)

The game $x \mid y$. Let x and y be numbers, and consider the game $x \mid y$. Then if $x < y$, this is the simplest number between x and y, so we shall consider the contrary case $x \geqslant y$. Then plainly the game $x \mid y$ has no dominated options. Moreover, its thermograph is the pyramid $\diagup x, y \diagdown$, and so $x|y$ determines the numbers x and y. It must therefore be in simplest form, for if the option y (say) were reversible, we should have $x \mid y = x \mid y^{LR}$ or $\{x \mid \}$, which games have different thermographs.

Now we assert that for any number z, we have $(x \mid y) + z = (x + z \mid y + z.)$ This is because in the difference

$$(x \mid y) + z + (-y - z) \mid (-x - z)$$

the moves not in z have exact counters, while the move for Right (say) from z to z^R is countered by Left's move to $-y - z$, since by the thermograph, we have $x \mid y > y + (z - z^R)$.

This kind of translation invariance allows us to normalise $x \mid y$ to the form $u \pm v$, where $u = \frac{1}{2}(x + y)$, $v = \frac{1}{2}(x - y)$. Of course it holds only for $x \geqslant y$, and shows that in this region, $x \mid y$ exhibits a strikingly continuous behaviour for all real numbers x and y.

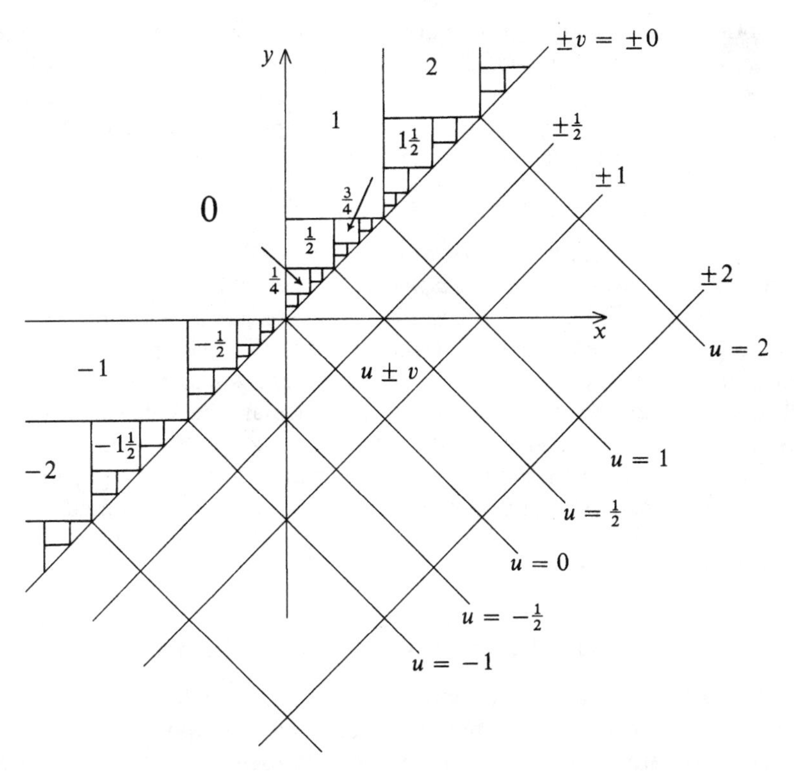

FIG. 18. The game $x \mid y$. Note: Points on boundaries here behave similarly to the points just South-East (↘) of them.

The game $x \pm a \pm b \pm \ldots \pm k$. Let $x, a, b, \ldots, k$ be numbers. Since $\pm t$ is its own negative, and is zero if $t < 0$, we can suppose that $a > b > c > \ldots > k \geqslant 0$. Then the thermograph of $\pm a \pm b$ is sketched in Fig. 19. This shows that $-a + b < \pm a \pm b < +a - b$. But this shows us that in the game

$$\pm a \pm b = \{a \pm b, \pm a + b \mid -a \pm b, \pm a - b\}$$

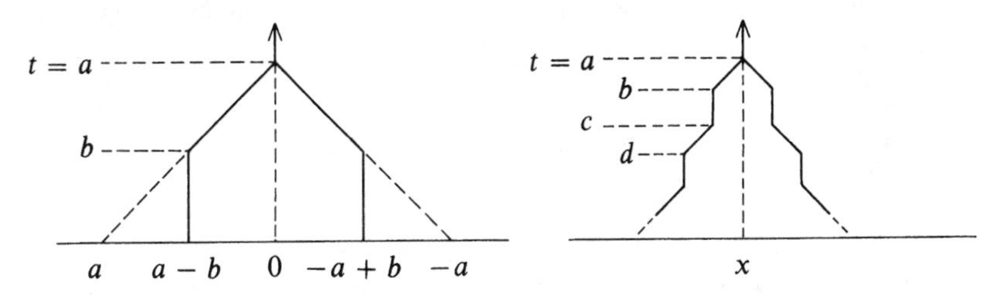

FIG. 19. Thermographs of $\pm a \pm b$ and $x \pm a \pm b \pm \ldots \pm k$.

the options $\pm a + b$ and $\pm a - b$ are dominated, for the difference between the two options on either side is just that between $\pm a \pm b$ and $a - b$. So in fact $\pm a \pm b = \{a \pm b \,|\, -a \pm b\}$, for since we know the simplest form $\{a + b \,|\, a - b\}$ of $a \pm b$, we can see that this option is not reversible. In a similar way, we find that the simplest form of $x \pm a \pm b \pm \ldots \pm k$ is

$$\{x + a \pm b \pm \ldots \pm k \,|\, x - a \pm b \pm \ldots \pm k\}$$

and that its thermograph is as shown.

Of course this uses the ordering $a > b > \ldots > k$, and is entirely concordant with experience and expectations. For since the game $\pm a$ represents an advantage of a move to the first player to move in it, when playing a sum of such games, the first player will take that with the largest a, then his opponent will take the next largest, and so on. In particular, the Left value will be $x + a - b + c - \ldots$, and the Right value $x - a + b - c + \ldots$.

In practice it is often simpler not to normalise games $x \,|\, y$ to the form $u \pm v$, but the rules still apply—in a sum of such games one should always move in that with the largest diameter $x - y$. (The diameter as here defined is twice the temperature of this game.)

DOMINO POSITIONS AND PROPOSITIONS

We return to the game with dominoes discussed in Chapter 7. To avoid pages full of little squares, we represent positions by graphs in which nodes represent squares, and edges join nodes representing adjacent squares. (Compare our conventions for COL and SNORT.) In this form, Left's move is to remove two nodes joined by a vertical edge, while Right removes a pair of nodes joined by a horizontal edge.

Note that the game could be played on any graph in which two kinds of edges are by definition called *horizontal* and *vertical*, but the addition of new such graphs does not seem to make the game any more interesting. Similar comments are often applicable to other games we shall discuss.

We attach at the end of the chapter a dictionary for dominoes like those for COL and SNORT. To show how the dictionary was prepared, we discuss in detail some of the results, and some particular positions. Most of the results referring to general positions are due to Norton.

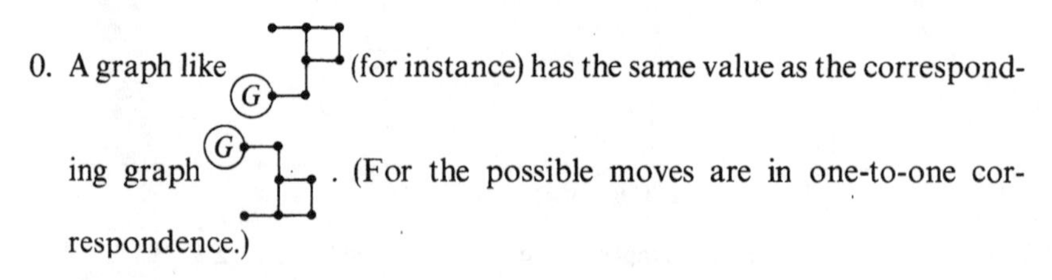

0. A graph like (for instance) has the same value as the corresponding graph . (For the possible moves are in one-to-one correspondence.)

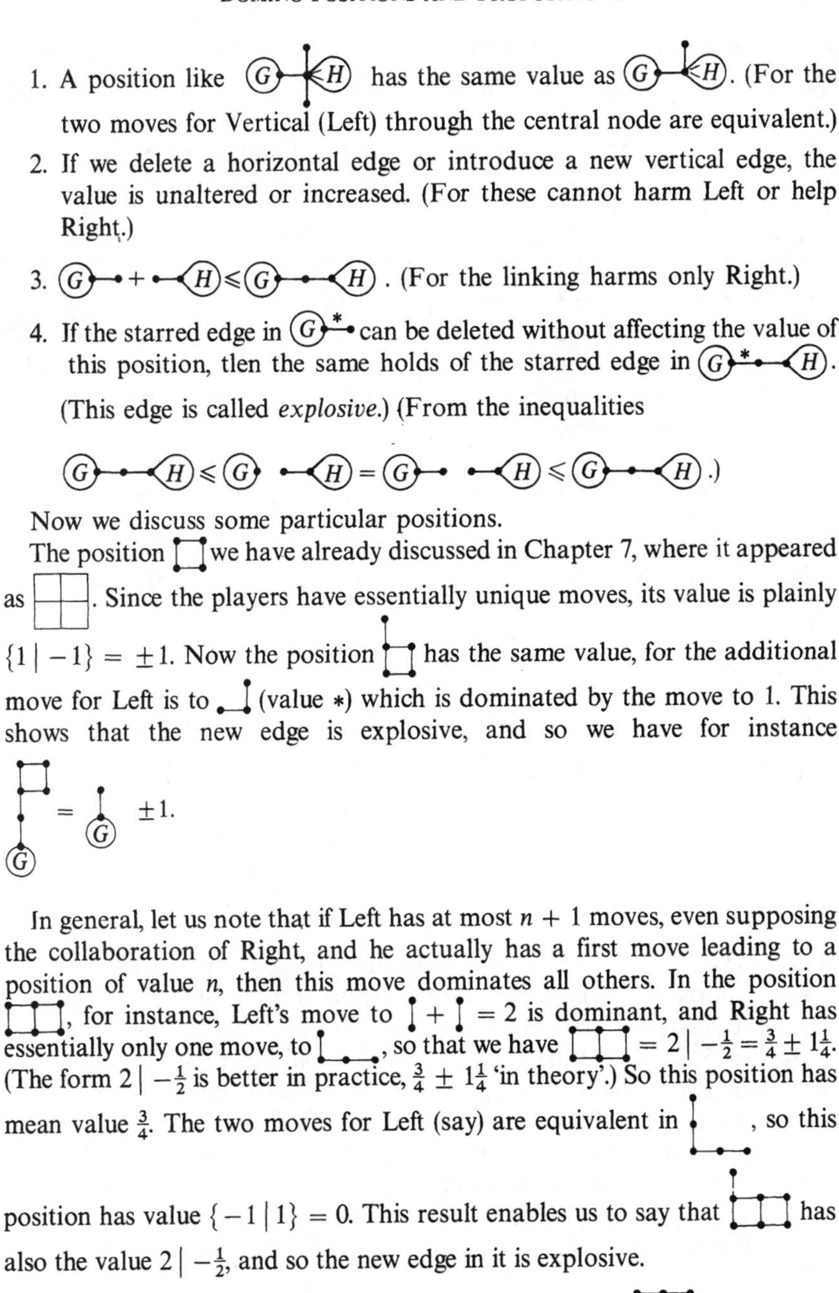

1. A position like (G)—(H) has the same value as (G)—(H). (For the two moves for Vertical (Left) through the central node are equivalent.)
2. If we delete a horizontal edge or introduce a new vertical edge, the value is unaltered or increased. (For these cannot harm Left or help Right.)
3. (G)— + —(H) ⩽ (G)——(H). (For the linking harms only Right.)
4. If the starred edge in (G)—* can be deleted without affecting the value of this position, tlen the same holds of the starred edge in (G)—*—(H). (This edge is called *explosive*.) (From the inequalities

(G)——(H) ⩽ (G) —(H) = (G)— —(H) ⩽ (G)——(H).)

Now we discuss some particular positions.

The position we have already discussed in Chapter 7, where it appeared as . Since the players have essentially unique moves, its value is plainly $\{1 \mid -1\} = \pm 1$. Now the position has the same value, for the additional move for Left is to (value $*$) which is dominated by the move to 1. This shows that the new edge is explosive, and so we have for instance

= (G) ± 1.

In general, let us note that if Left has at most $n+1$ moves, even supposing the collaboration of Right, and he actually has a first move leading to a position of value n, then this move dominates all others. In the position , for instance, Left's move to $+$ $= 2$ is dominant, and Right has essentially only one move, to , so that we have $= 2 \mid -\frac{1}{2} = \frac{3}{4} \pm 1\frac{1}{4}$. (The form $2 \mid -\frac{1}{2}$ is better in practice, $\frac{3}{4} \pm 1\frac{1}{4}$ 'in theory'.) So this position has mean value $\frac{3}{4}$. The two moves for Left (say) are equivalent in , so this position has value $\{-1 \mid 1\} = 0$. This result enables us to say that has also the value $2 \mid -\frac{1}{2}$, and so the new edge in it is explosive.

We are now in a position to evaluate the 3×3 square . (We should obviously describe sizes in terms of nodes here, since these correspond to squares in the original game.) In the position 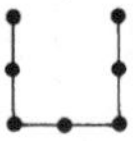 , the two moves for Left are to  (0) and  ($\frac{1}{2}$), while that for Right is to  +  (2). So  $= \frac{1}{2} \mid 2 = 1$. Now we have the equation

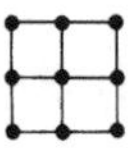

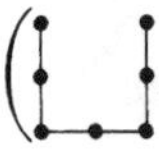

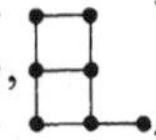

$$= \pm \left(\; , \; \right) = \{1, \tfrac{1}{2} \mid -2 \,\|\, -1, 2 \mid -\tfrac{1}{2}\} = G \text{ say.}$$

Now it is trivial that $-2 \leqslant G$ (add 2 to G and see how easy it is to win), so the Left option $\frac{1}{2} \mid -2$ is reversible through -2, and so can be replaced by the Left options (there aren't any) of -2. In this way, we see that G simplifies to ± 1, its simplest form.

It is not hard to show that the 3×4 rectangle 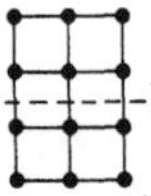 has value $1\frac{1}{2}$. For this is plainly a lower bound (break the rectangle across the dotted line), and a quick strategic discussion shows that Left cannot win the difference

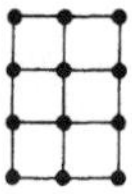

$$- 1\tfrac{1}{2}.$$

Larger rectangles are something of a problem. But if we only want to work out who wins, we can employ the following type of argument. From the 4×4 square, Left can move to

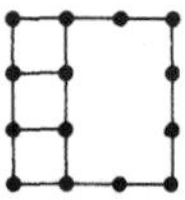

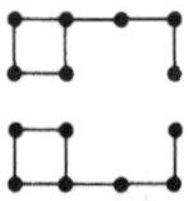

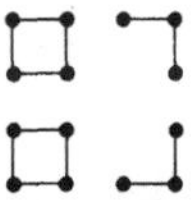

$$\geqslant \; = \; = 0,$$

and so the 4×4 square is a win for the first player. Similar arguments can be found for the 4×6 rectangle, using the value of the position 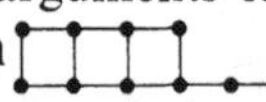 , which does not take too long to compute.

The 5×5 square can be shown to be a second player win (and so have value 0) by the following special strategy. This gives Left 6 moves, *or* keeps Right down to 5 moves and gives Left 5 moves.

Supposing Right's first move is in the top left 3×3 square, make the moves a, b of the first drawing if we can, followed by any c in the top three rows, and the moves d, e, f. If not, make the move b of the later drawings and occupy the centre if possible, followed by any move d *other* than e and f of the second drawing, then such of those moves e and f which are still legal. If this is impossible, the move c of the third drawing might be available, and lead back to a similar strategy to our first attempt.

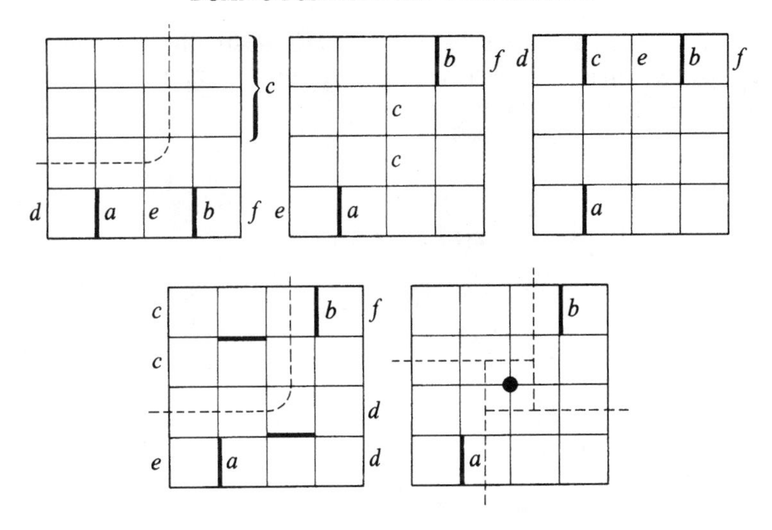

FIG. 20. The equation $5 \times 5 = 0$.

If not, the position is *either* as in the fourth drawing, with Right's third move in the top left 3×3 square, say, when we can make the moves of that drawing, *or* Right has taken the centre. In the last case, we do not really break any horizontal lines when we partition as indicated into 4 regions of values $1, \frac{1}{2}, -\frac{1}{2}, -\frac{1}{2}$, and so we are at least half a move ahead.

The zig-zag patterns $\cdot$, , , , , ... give rise to an interesting sequence of values. Letting ZZ_n be the value of the n-square zig-zag pattern, we find $ZZ_1 = 0$, $ZZ_2 = 1$, $ZZ_3 = *$, $ZZ_4 = 1 \mid 0$, $ZZ_5 = \pm 1$, $ZZ_6 = 2 \mid *$, $ZZ_7 = \pm 1*$, $ZZ_8 = 2 \mid 1 \parallel 0$, $ZZ_9 = \pm(2 \mid 0, 2 \mid *)$, $ZZ_{10} = 1*$, and $ZZ_{11} = ZZ_9*$. The later values get more complicated, but we can fairly easily calculate them *almost exactly*.

In fact we find

$$ZZ_{8n+1 \text{ or } 8n+3} = 0\text{-ish}$$

$$ZZ_{8n+1 \text{ or } 8n-3} = \pm 1\text{-ish}$$

$$ZZ_{8n+2} = 1\text{-ish}$$

$$ZZ_{8n-2} = 2 \mid 0\text{-ish}$$

$$ZZ_{4n} = (n \mid n-1 \parallel n-2 \mid\mid\mid n-3 \ldots 1 \mid\mid\mid\mid\mid\mid\mid\mid\mid 0)\text{-ish}$$

where the suffix "-ish" means "infinitesimally shifted". In other words, we write G-ish for $G + \varepsilon$, when ε is infinitesimal. In these particular cases, of course, the various infinitesimal shifts ε are small games.

The game ZZ_{4n} has mean value $1 - (1/2^n)$, is strictly less than 1, and strictly greater than any negative number, but not greater than 0. These results follow from the thermographs:

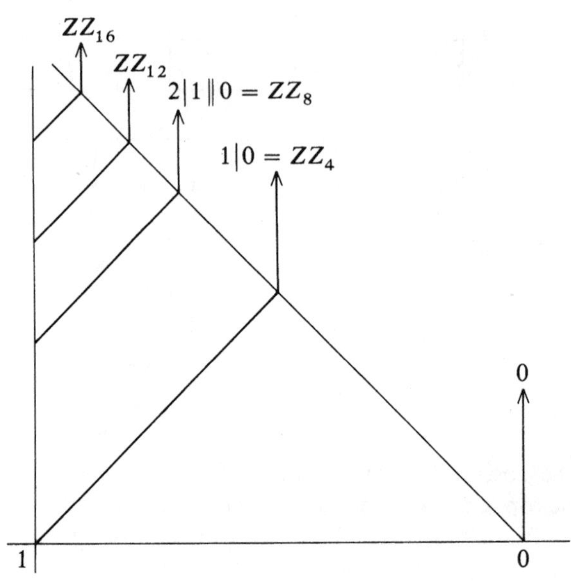

FIG. 21.

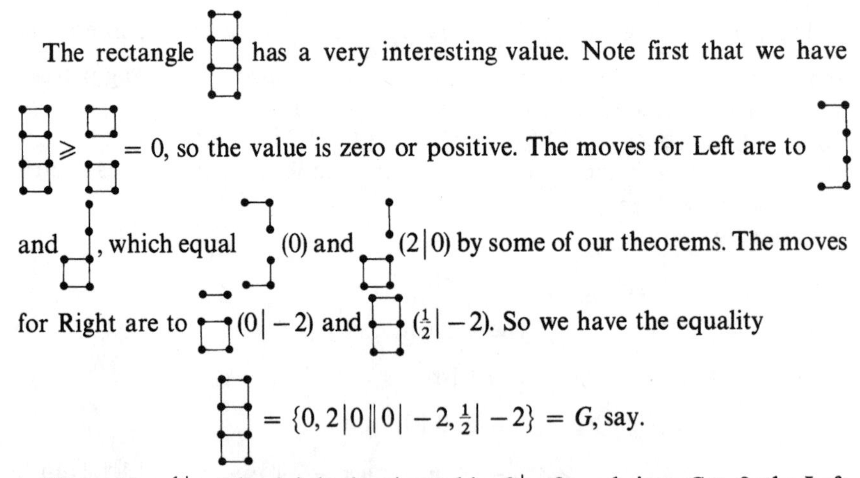

The rectangle [figure] has a very interesting value. Note first that we have [figure] $\geqslant$ [figure] $= 0$, so the value is zero or positive. The moves for Left are to [figure] and [figure], which equal [figure] (0) and [figure] $(2|0)$ by some of our theorems. The moves for Right are to [figure] $(0|-2)$ and [figure] $(\frac{1}{2}|-2)$. So we have the equality

$$[\text{figure}] = \{0, 2|0 \| 0|-2, \tfrac{1}{2}|-2\} = G, \text{ say.}$$

The option $\frac{1}{2}|-2$ is plainly dominated by $0|-2$, and since $G \geqslant 0$, the Left option $2|0$ is reversible to 0, and can be replaced by the (non-existent) 0^L. So we have $G = 0 \| 0 | -2$, which, since we see it is strictly positive, is in simplest form.

For reasons that will only appear later, this game is called $+_2$ (pronounced "tiny-two"). For any positive number x, we have a similar game "tiny-x"

$$\text{"tiny-}x\text{"} = 0 \| 0 | -x = +_x.$$

For each positive x, $+_x$ is a positive infinitesimal, and indeed a *small* game in the sense of Chapter 9, since it is strictly smaller than all positive numbers. But other calculations show that for strictly positive x these games are smaller than all positive all small games, such as $\uparrow$. As a matter of notation, we abbreviate sums involving such games in a natural way—thus $5+_2$ means $5 + +_2$, and $5 -_2$ means $5 - +_2$.

It is possible to define powers $\uparrow^x$ of $\uparrow$ for positive $x \geqslant 1$ so that whenever $x > y$, then $\uparrow^x$ is infinitesimal compared to $\uparrow^y$, and all these powers are all small. We have thus a rough-and-ready scale of infinitesimals:

Firstly, infinitesimal numbers, like $1/\omega$, $1/\varepsilon_0$, etc.

Next, the all small games, such as $\uparrow$, $\uparrow^2$, etc.

Finally, the games like $+_1$, $+_2$, etc.

We say *finally* because indeed the games $+_\alpha$ really tend to zero as α tends to **On**, any strictly positive game being bigger than some $+_\alpha$. (Any *short* positive game is greater than some $+_n$.) But we should also add that, *zerothly*, there are some infinitesimal games that are strictly greater than all infinitesimal numbers! These remarks are very much amplified in Chapter 16.

A DOMINO DICTIONARY

We now tabulate values for all domino positions with at most 6 nodes (Fig. 22). The game of dominoes has a behaviour in some ways intermediate between the two games COL and SNORT of this chapter, with typical values not so restricted as those of COL nor so chaotic as those of SNORT. Many of these are derivable from each other by simple rules. Often it obviously does not affect play if we make a configuration bend in the opposite direction to the given one—for instance = . There are also a number of rules telling us that on certain occasions edges may be deleted without affecting values, as described earlier in this chapter. Here is a brief catalogue of explosive edges:

The ones indicated by lightning bolts in:

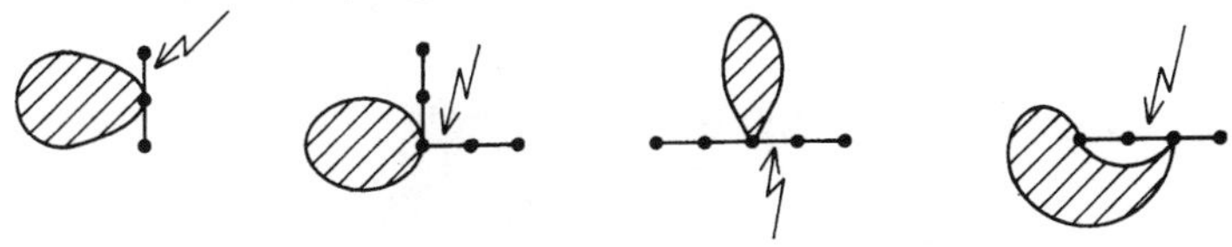

FIG. 22.

and edges joining any one of the configurations below to any one of the indicated surrounding nodes:

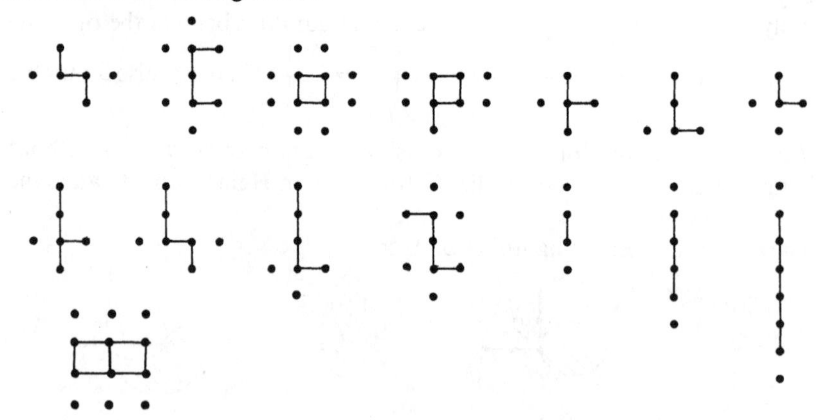

Some fairly large domino positions we have analysed are:

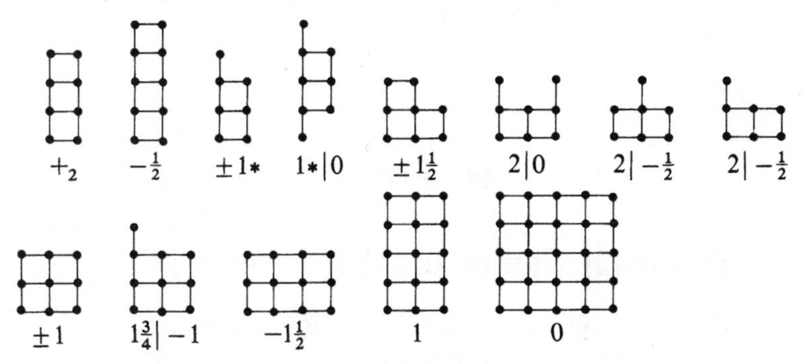

We have chosen these as being of shapes fairly likely to arise in actual play.

Göran Andersson's Domino game is called "Domineering" in *Winning Ways*, where larger dictionaries can be found. The necessary evaluations have been greatly eased by David Wolfe's "combinatorial games toolkit" for partizan game theory which runs on Linux computers. You can obtain it from

http://www.gustavus.edu/~wolfe/papers-games/

or by sending e-mail to wolfe@gustavus.edu.

CHAPTER 11

Impartial Games and the Game of Nim

CORPORAL NYM: *I have operations in my head, which be humours of revenge.*

William Shakespeare, The Merry Wives of Windsor

This chapter is intended to show how the Sprague–Grundy theory of impartial games fits into our more general ideas. The theory will itself be developed inside the chapter.

Definition. The game G is *impartial* if and only if for every position $P = \{L \mid R\}$ of G, we have $L = R$ (as sets).

In other words, G is impartial only if every option of G is also impartial, and the collection of Left options coincides with the collection of Right ones.

Recall the convention $\{A, B, C, \ldots\} = \{A, B, C, \ldots \mid A, B, C, \ldots\}$. In view of this convention, it is natural to use G', rather than G^L or G^R, for the typical option of G, and to write $G' \in G$ to mean that G' is an option of G. So we identify each game with the set of all its options.

THE GAME OF NIM

This game is played with a number of heaps of matchsticks. The legal move is to strictly decrease the number of matchsticks in any heap (and throw away the removed sticks). A player unable to move because no sticks remain is the loser.

It is obvious that Nim is the disjunctive sum of its heaps. So we can analyse it by writing $*n$ for the value of a heap size n. Inductively, these *impartial numbers*, or *Nim-numbers* are defined by

$$*n = \{*0, *1, \ldots, *(n-1)\} = \{*m\}_{m<n}.$$

We note in particular the values

$$*0 = \{\,|\,\} = 0, *1 = \{0 \mid 0\} = *, \text{ and}$$

$$*2 = \{0, * \mid 0, *\}.$$

(We shall continue to use the abbreviations 0 and $*$.)

Now without assuming the general theory of earlier chapters, we shall develop the Sprague–Grundy theory in an analogous but easier way. The idea is perhaps best illustrated by reference to another game.

THE SILVER DOLLAR GAME, WITHOUT THE DOLLAR

This game is played on a semi-infinite strip of squares, with a finite number of coins, no one of which is a Silver Dollar. Each coin is placed on a separate square, and the legal move is to move some coin leftwards (i.e. towards the finite end of the strip), not passing over any other coin, onto any unoccupied square. The game ends when some player has no legal move, because the coins are in a traffic jam at the left end of the strip.

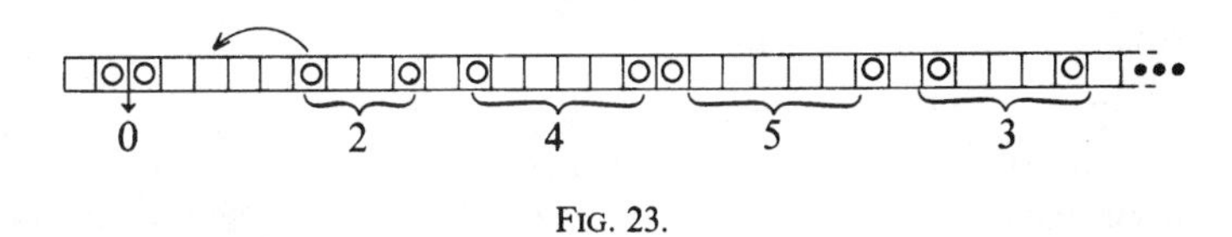

FIG. 23.

Figure 23 illustrates a typical position and a typical legal move. (Of course all games are impartial in this Chapter, so the move is legal for either player.) Now we assert that this game is merely a disguised and slightly generalised form of Nim.

Here is the disguise revealed. Starting from the rightmost coin, count the numbers of squares in alternate spaces between the coins, and let these numbers be the sizes of Nim-heaps. So the illustrated position corresponds to the Nim-position 3, 5, 4, 2, 0.

Now we assert that despite certain differences, which are somewhat startling, this game really does behave like Nim. Notice first that every move in the coin game affects just one of our numbers, just as every move in Nim affects just one of the heaps. Observe also that there are moves in the coin game which decrease any one of the numbers by an desired extent. So the only apparent difference is that in the coin game there are sometimes moves which *increase* one of the numbers—for instance the indicated move would increase 2 to 5.

However, these increases are not needed by the winning player, and they are of no avail to his opponent. For if I am winning, and you increase 2 to 5 (say), then I can plainly respond by simply decreasing 5 to 2 again. In the game above, I shall simply follow your move by moving the coin just right of yours a corresponding three places.

The argument is perfectly general and proves the following theorem.

THEOREM 70. *Let G be any game played with a finite collection of numbers (from* 0, 1, 2, 3, ...) *in the following way. Each move affects just one number, and strictly changes that number. Any decrease of a number is always obtainable by a legal move, but some increases may also be possible. However, the rules of the game are such as to insure that it always terminates. Then the outcome of any position in G is the same as that of the corresponding position in Nim.*

Proof. The player who has the winning strategy in Nim need not make use of the new moves. If his opponent does, he can always move so as to restore the status quo, and the rules ensure that this brings us nearer to the end of the game.

In the terminology of Chapter 10, the increasing moves are *reversible.* This result immediately gives us *Grundy's theorem;*

THEOREM 71. *Each short impartial game G is equivalent in play to some Nim-heap.*

Proof. Suppose that this is true of all the options $A, B, C, \ldots$ of G, so that these positions are equivalent to Nim-heaps of sizes $a, b, c, \ldots$, say. Now let n be the *least number* (from $0, 1, 2, 3, \ldots$) *which does not appear among the numbers* $a, b, c, \ldots$. This number is the *mex* (minimal excludent) of $a, b, c, \ldots$. We assert that G is essentially a Nim-heap of size n. For certainly all the numbers $0, 1, \ldots$ which are *less* than n must appear among the numbers $a, b, c, \ldots$, so that any *decrease* of n is obtainable by some legal move. Perhaps some increases are possible (if one of $a, b, c, \ldots$ exceeds n), but it is certainly not possible to move to n itself. So in the sense of Theorem 70, G behaves like a Nim-heap of size n.

Note. This proves that the value of any impartial short game is one of the impartial numbers $0, *, *2, *3, \ldots$. A purely formal inductive proof could also be given, and indeed the theorem follows almost instantly from Theorem 69.

INFINITE NIM

We can generalise Nim by allowing the sizes of the heaps to be arbitrary ordinals α, the legal move being to replace any α by a strictly smaller ordinal β. There are therefore impartial numbers for all ordinals, defined by

$$*\alpha = \{*\beta\}_{\beta<\alpha}.$$

Theorem 71 generalises to show that *every* impartial game is equivalent to some $*\alpha$.

In these theorems we have for clarity used the word *equivalent* where in most parts of the book we should simply write *equal*. We repeat our conclusions once again:

If the options of G are equal in value to certain impartial numbers $*a$, $*b$, $*c$, ... then provided G itself is impartial, it is equal to the impartial (ordinal) number $*n$, where n is the least number not appearing among the numbers $a, b, c, \ldots$. The number n is usually called the *Grundy number* of G. Our treatment is different from that of Grundy, and we must point out that Sprague had earlier discovered the theory independently of Grundy, and in a still different way.

Now the benefit of this approach is that we see that the game of Nim itself *must* have a solution of a certain kind, even before we can see what the exact details are. For since the disjunctive sum

$$*a + *b$$

is itself an impartial game, it must have a Grundy number, n say, where n is some function of a and b, so that we shall have

$$*a + *b = *n.$$

The theory of Nim will follow as soon as we have computed exactly what function n is of a and b.

This we can do inductively if we like, using the definition of the disjunctive sum. This tells us that $*a + *b = *n$, where n is the least number not the Grundy number of any of the sums

$$*a' + *b, \qquad *a + *b' \qquad (a' < a, b' < b).$$

From this it is easy to compute $*a + *b$ recursively, and in fact of course we have already done so in Chapter 6, where the reader will find a table for all $a < 16$, $b < 16$.

Since when playing games it is handy to have Nim-sums at one's fingertips, we display all cases with numbers less than 8 in Fig. 24. The lines of this diagram represent triples of numbers any two of which Nim-add to the third.

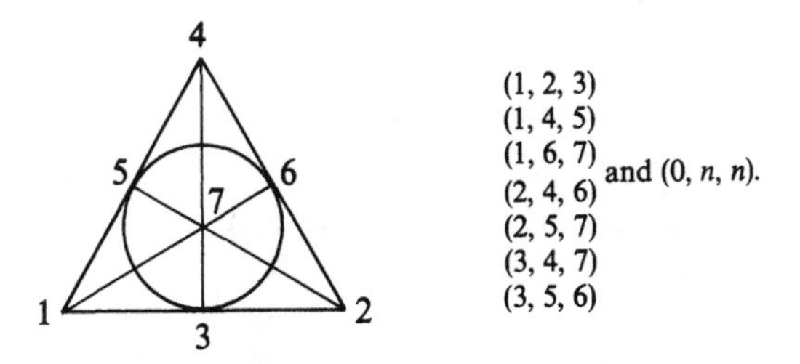

FIG. 24. Some Nim-triplets.

So for instance the circular "line" tells us that $3 +_2 5 = 6$, $3 +_2 6 = 5$, $5 +_2 6 = 3$. (We use $+_2$ for the Nim sum, and read $3 +_2 5$ as "three Nim five".) Many readers will find themselves able to memorise these Nim-triplets without the diagram—we give a list beside it. In general we have the triplets $(1, 2n, 2n + 1)$, $(7, n, 7 - n)_{n<7}$, and we can replace 7 in the latter by any number $2^k - 1$. This, together with the special triplet (3, 5, 6), and occasionally $(8, n, 8 + n)_{n<8}$, is all one should ever need.

To find a good move from a general Nim-position, the first step is to compute its Nim-sum. If this is zero, the position is a second player win, so your best hope is to leave the position as complicated as possible so that your opponent will fail to analyse it. But if the Nim-sum is non-zero, we can Nim-add the sum to at least one of the heap-sizes in such a way as to cause a decrease, and this determines a legal move to a position of Nim-sum zero.

So for instance in the position 3, 4, 8, 9 the Nim-sum is

$$3 +_2 4 +_2 8 +_2 9 = 3 +_2 4 +_2 1 = 3 +_2 5 = 6,$$

which is non-zero. Nim-adding 6 to the numbers 3, 4, 8, 9 we find 5, 2, 14, 15 respectively, and so the only good move is to decrease 4 to 2. In practical play one should try to visualise the matchsticks in each heap partitioned into distinct parts whose sizes are powers of 2, and then a good move is often obvious. For instance in Fig. 25 when we partitition the heaps (mentally) as indicated, it is obvious that reducing the second heap from 4 to 2 will "cure" the position.

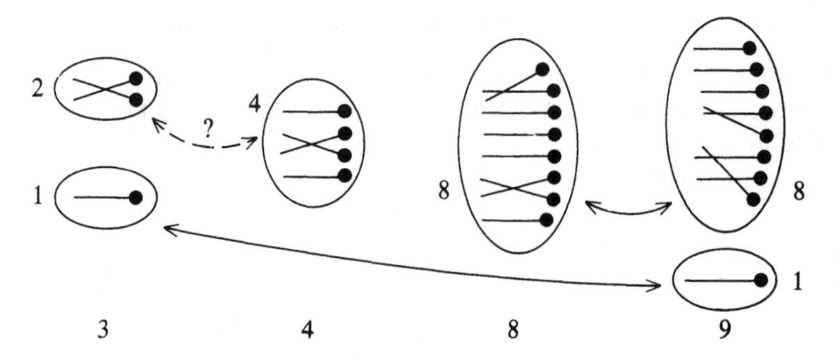

FIG. 25. A move made plain.

One should also get into the habit of realising that once one has evaluated the Nim-sum of a position, one has really proved that it is equivalent to a Nim-heap of a certain size. In particular, for instance, if some sub-position has value zero, it can be neglected until such time as our opponent moves in it, when we respond by reversing it to zero again. But more generally, any sub-position of value 3, say, may and should be thought of as a disguised Nim-heap of three sticks.

THE GAME OF KAYLES

This was introduced by Dudeney and Loyd. It is played by skilful players with a number of rows of ninepins. See Fig. 26.

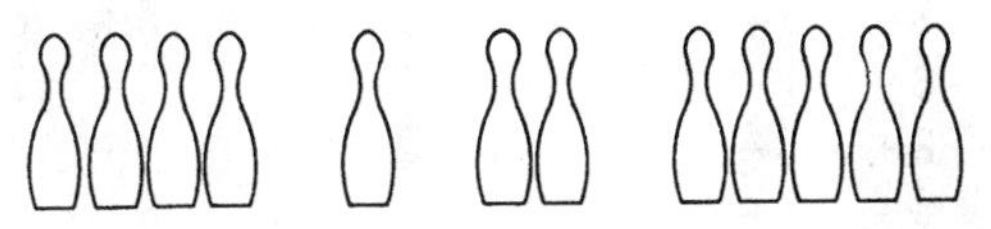

FIG. 26. The Kayles position $K_4 + K_1 + K_2 + K_5$.

We suppose the players are so skilful that they can throw a ball so as to knock down any single ninepin or any two adjacent ones, but that it is physically impossible to knock down pins separated by any larger distance. The last mover wins.

Let us write K_n for the value of a row of n pins in Kayles. Plainly any Kayles position is a disjunctive sum of its rows. The legal moves from K_n are to sums $K_a + K_b$, where a and b are restricted only by the conditions $a \geqslant 0$, $b \geqslant 0$, $a + b = n - 1$ or $n - 2$.

So we have

$$K_0 = \{\ \} = 0\,(= *0)$$

$$K_1 = \{K_0\} = \{0\} = *1\,(= *)$$

$$K_2 = \{K_0, K_1\} = \{0, *1\} = *2$$

$$K_3 = \{K_1, K_2, K_1 + K_1\} = \{*1, *2, *1 + *1\} = \{*1, *2, 0\} = *3$$

$$K_4 = \{K_2, K_1 + K_1, K_3, K_2 + K_1\} = \{*2, 0, *3, *2 + *1\}$$

$$= \{*2, 0, *3\} = *1.$$

Note that K_4 is not given by the *next larger* number than 2, 0, 3, but the *least absent* number, namely 1. Continuing, we find

$$K_5 = \{K_3, K_2 + K_1, K_4, K_3 + K_1, K_2 + K_2\} = \{*3, *3, *1, *2, *0\} = *4,$$

and then $K_6 = *3$, $K_7 = *2$, $K_8 = *1$.

In the standard language, the Grundy numbers of the positions

$$K_1,\ K_2,\ K_3,\ K_4,\ K_5,\ K_6,\ K_7,\ K_8$$

are

$$1,\ \ 2,\ \ 3,\ \ 1,\ \ 4,\ \ 3,\ \ 2,\ \ 1,$$

respectively.

We believe that the complete K-series was first calculated by Richard Guy, who discovered the astonishing fact that K_n is a periodic function of n with period 12, for all $n \geqslant 72$. (The same fact has been independently discovered by a number of other people.) We tabulate the K-series from $n = 0$ in rows of 12 to illuminate the periodicity:

0+	1+	2+	3+	1−	4+	3−	2−	1−	4+	2+	6+
4−	1+	2+	7−	1−	4+	3−	2−	1−	4+	6−	7−
4−	1+	2+	8+	5+	4+	7+	2−	1−	8+	6−	7−
4−	1+	2+	3+	1−	4+	7+	2−	1−	8+	2+	7−
4−	1+	2+	8+	1−	4+	7+	2−	1−	4+	2+	7−
4−	1+	2+	8+	1−	4+	7+	2−	1−	8+	6−	7−
4−	1+	2+	8+	1−	4+	7+	2−	1−	8+	2+	7−

Grundy numbers for Kayles, from $n = 0$ to $n = 83$

Here the values are to be read straight across the rows, and the last row now repeats indefinitely. The signs "+" and "−" are to be ignored for the moment.

OTHER IMPARTIAL GAMES

The Grundy numbers for many other games have been shown by Guy, C. A. B. Smith, and others, to exhibit similar behaviour. Often there is "almost" periodicity present from the very beginning, which later may or may not "settle down" into exact periodicity. In other cases there is no real evidence of any kind of periodicity, although no octal game has been definitely shown not to be ultimately periodic.

These *octal* games generalise both Nim (the case ·3333...) and Kayles (the case ·77). In general we have a game $\cdot A_1 A_2 A_3 \ldots$ for any sequence of "digits" $A_1, A_2, A_3, \ldots \geqslant 0$. If the digit A_k has the binary expansion $2^a + 2^b + \ldots$, this means that it is legal to remove just k objects from any heap and then partition the remainder of that heap into a number a or b or ... (only) of non-empty heaps. Such moves, as k varies, are the only legal moves.

Thus, since $3 = 2^1 + 2^0$, in the game ·333..., it is permissible to remove any positive number k of counters from a heap, leaving the remaining ones (if any) to form either 1 heap or 0 heaps. This game is therefore Nim. In the case ·77, we may remove only 1 or 2 objects from a "heap", leaving the remaining ones forming 0 or 1 or 2 heaps (since $7 = 2^0 + 2^1 + 2^2$). So we might as well think of the objects as arranged in a line and remove 1 or 2 adjacent ones, as in Kayles.

As a more general example, we take ·156. Thinking of the heaps as lines again, we see that (since $1 = 2^0$) we can remove a single object if and only if it forms an entire line, two adjacent objects if and only if they form an entire line *or* are strictly inside a line (since $5 = 2^0 + 2^2$), and three adjacent objects if and only if they do not form an entire line (since $6 = 2^2 + 2^1$). We consider this example because J. C. Kenyon has found that its Grundy numbers are periodic with period 349 from 3479 onwards!

The octal games are those games $\cdot A_1 A_2 \ldots$ in which each $A_k < 8$, and they form a fairly natural class since they have easy interpretations in terms of lines of objects. But digits $\geqslant 8$ are meaningful, and have also been considered. Another extension is to allow certain digits $\ldots A_{-1} A_0$ *before* the point, provided suitable conditions are satisfied. Thus 4·33 denotes the game in which any heap may be split into two non-empty parts ($2^2 = 4$), or reduced by 1 or 2 objects.

We shall not discuss these games in detail—for a more comprehensive treatment see *Winning Ways* and the references given therein. But we cannot resist noting Guy's beautiful discovery that the game $\cdot 7^{2m}$ ($m = 2^n$) has for its Grundy number sequence the sequence obtained from the ordinary Kayles sequence by replacing each table entry $x+$ by the sequence

$$mx, mx + 1, \ldots, mx + m - 1,$$

and each entry $x-$ by the same sequence reversed. Nor can we resist pointing out that Berlekamp's remarkable theory of the schoolboy game of *Dots and Boxes* shows that one must understand the theory of Kayles to become an expert at this game. See *Winning Ways*, and Berlekamp's book *The Dots-and-Boxes Game: Sophisticated Child's Play*.

The Grundy number series for Grundy's own game (split any heap into two non-empty heaps of distinct sizes) has now been analysed for $n < 10^5$ without discerning any permanent periodicity. There is a most remarkable initial tendency towards the period three, but the permanence of this or any of the other "almost periods" seems doubtful. We shall discuss Grundy's game again in Chapter 12, where we shall disprove a conjecture about the misère form of the game.

For games in which the typical position depends on just one parameter n, the Grundy theory is essentially complete—all we need to play the game is a table of (or formula for) the Grundy number of the nth position. We give some examples not exactly of the octal type:

PRIM (remove from a heap of size n any number prime to n—invented by Alan Tritter)

$n =$	1	2	3	4	5	6	7	8	9	10	11	12	13	14	15...
G-series:	0	1	2	1	3	1	4	1	2	1	5	1	6	1	2...

In general $G(n)$ is k if the least prime divisor of n is the kth prime. If we allow the removal of 1 from 1, then the G-values 0 and 1 are interchanged.

DIM (remove a divisor of n from a heap of size n)

$$n = 0\ 1\ 2\ 3\ 4\ 5\ 6\ 7\ 8\ 9\ 10\ 11\ 12\ 13\ 14\ 15$$

$$G\text{-series: } 0\ 1\ 2\ 1\ 3\ 1\ 2\ 1\ 4\ 1\ \ 2\ \ 1\ \ 3\ \ 1\ \ 2\ \ 1\ldots$$

In general $G(n) = k$ if 2^{k-1} exactly divides n. If we disallow the removal of n from n the G-values are decreased by 1.

More complicated games are tackled as usual by building a dictionary of small positions and looking for some general patterns. Even when, as usual, no complete theory emerges, we usually find enough to enable us to play the game against intelligent opponents ignorant of the Grundy theory and win almost every time. We recommend the reader who wants to try his hand to tackle the game of dominoes on a Chessboard, when we allow each player to put his dominoes in either the horizontal or vertical orientations.

The game of *Hackenbush unrestrained* should also be mentioned here—it is played exactly like the variety of Hackenbush described in Chapter 8, except that there is just one kind of edge, and each edge may be chopped by either player. We give a complete discussion in Chapter 13, since the theory really involves the *animating functions* discussed there, but some readers will prefer to try things for themselves. (The *theory* is considerably easier for *trees* than for general pictures, although the *answers* in the general case are almost as easy to guess.)

THE SILVER DOLLAR GAME, WITH THE DOLLAR

This game is played just like the corresponding game without the Dollar, except that just one of the coins we use is a Silver Dollar, and the leftmost square is replaced by a moneybag, capable of holding any number of coins. So the leftmost coin on any square other than the moneybag may if we like be put into the moneybag as a move. When the Dollar is in the bag, the game ends, and the person who did *not* put the Dollar into the bag pockets the bag and goes home.

The theory is exactly the same as in the simpler game, except that the moneybag counts as an empty square if the next coin to the right of it is anything other than the Silver Dollar, but a full square when it is the Dollar. (Because we don't want to put the Dollar into the bag we prefer to think of it as full when the Dollar is the nearest coin to it!) Since in Nim we are never forced to make any heap have size -1, we shall never be forced to put the Dollar in the bag, if we can win the Nim game.

If instead the person who puts the Dollar into the bag may pocket the bag as part of the same move, the coin we don't want to put into the bag becomes the one to the left of the Dollar. So in this case we count the bag as full only when it is this coin which is next to the right of it. The theory is otherwise unaltered. I believe the Silver Dollar game is due to N. G. de Bruijn.

NORTHCOTT'S GAME

This game is played with the pawns on a Chessboard, with positions like those of Fig. 27 in which each row contains one black and one white pawn. The pawns may move freely (many squares at a time) along the rows, but may not jump over each other. A player loses when unable to move.

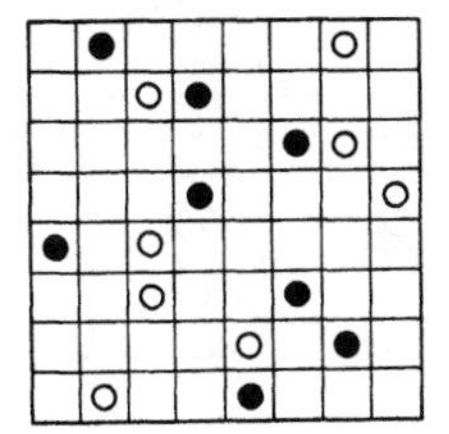

FIG. 27.

Despite the potential infinitude of the game, it is really Nim, played on the numbers of spaces between opposing pawns. Thus the position of Fig. 27 has Grundy number $4 + 0 + 0 + 3 + 1 + 2 + 1 + 2 = 7$, the addition being Nim-addition. The winning player should always "close in" on his opponent, whose attempts to retreat will then be unavailing.

Another variant of Nim, which some will find more appealing than the original, is played with spots on a piece of paper. The *Rims* move is to draw a closed loop passing through any positive number of spots but not meeting

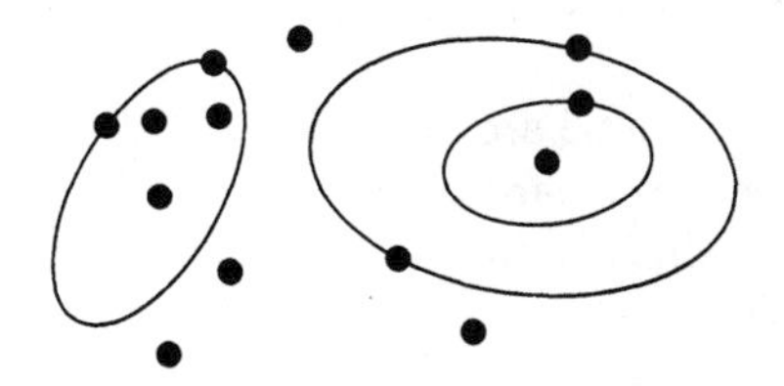

FIG. 28. A position in Rims (or in Rayles).

any other loop. So Fig. 28 shows the Nim-position with heaps of sizes 3, 4, 0, 1. We obtain *Rayles* if we insist that each loop pass through just one or two spots (a reformulation of Kayles). The other octal games can also be reformulated in this way.

DIMINISHING RECTANGLES

This game is played with a number of rectangular cards placed on a table, and a large bin containing an indefinitely large stock of such cards. Each card measures an integral number of inches in each direction. The move is rather curious—we take an $a \times b$ card from the table and an $A \times B$ card from the bin, and cut the $A \times B$ card once in each direction so as to remove an $a \times b$ card from one corner, and three other cards of sizes $a' \times b$, $a \times b'$, and $a' \times b'$, say. Then provided that $a' < a$ and $b' < b$, we may replace the $a \times b$ card originally on the table by these three new cards, throwing the two $a \times b$ cards into the bin. See Fig. 29.

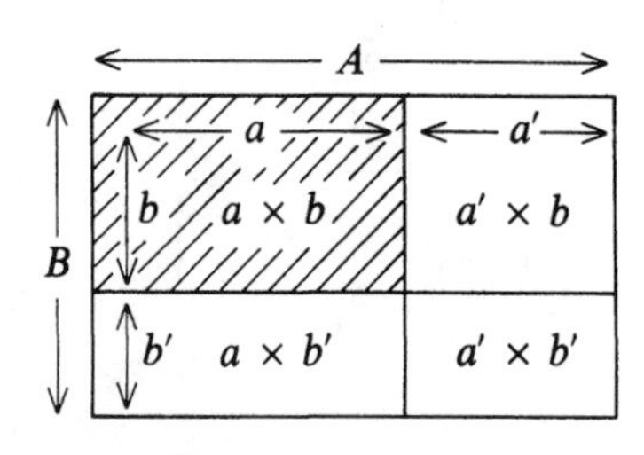

FIG. 29.

In other words we may replace a table card by the three pieces left when it is cut from a bin card, *provided* that these are all strictly smaller than the table card. The game ends as usual when the table is empty, so that no player has a legal move. What is the strategy?

Of course this curious game has been designed to make use of the curious theory developed in Chapter 6. Each rectangle has an *area* defined as the product of the lengths of its edges in the sense of Chapter 6, and the cards on the table have then a total area obtained by summing these areas in the sense of that Chapter (and this). The correct moves are to positions of total area zero.

So if the cards on the table are of sizes 1×1, 2×2, 4×4, 8×8 a good move is to replace the 8×8 card by three cards of sizes 8×7, 7×8, 7×7, whose total area is 4, which is also the total area $1^2 + 2^2 + 4^2 = 1 + 3 + 6$ of the remaining cards.

THE DELIAN PROBLEM RESOLVED

We can generalise this game in the obvious way to cards whose edges have arbitrary ordinal lengths, and which have arbitrarily many dimensions. In the generalised game, for what n is the position $2 \times 1 \times 1 + n \times n \times n$ a win for the second player?

This question of course reduces to solving the equation $n^3 = 2$ in the sense of Chapter 6, or in other words to the Nim-duplication of the cube. The answer is contained in that Chapter, the simplest value of n being the least infinite ordinal ω. See Fig. 30.

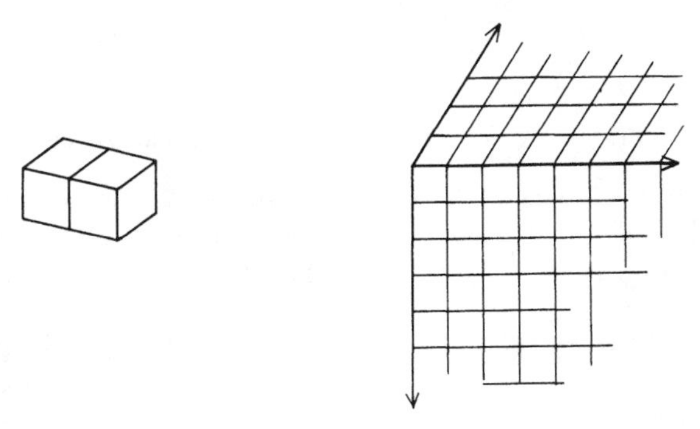

FIG. 30. Two cuboids of the same volume!

The solution sketched in this figure is not unique, since $n = 2\omega$ and 3ω are also solutions, 2 and 3 being the non-trivial cube roots of 1.

Distinctly more natural games involving Nim-multiplication are the coin-turning games of H. W. Lenstra, Jr., described in Chapter 14 of *Winning Ways*.

THE SMITH THEORY FOR GAMES WITH INFINITE PLAY′

C. A. B. Smith has extended the Grundy–Sprague theory to cover games in which the play need not terminate (as in Northcott's game above, when played badly). If the play continues forever, we call the game a draw. We give Smith's theory here, with an informal, though perfectly rigorous, proof.

We draw the *graph* of the game, which may be finite or infinite, having a node for each position and a direct edge from P to Q when it is legal to move from P to Q. (Of course, we are considering only impartial games.) We are allowed to *mark* a position P in this graph with the number n (for infinite graphs, n may be an arbitrary ordinal) in the following circumstances. Firstly, n must be the mex of all the numbers that already appear as marks of any of the options of P. Secondly, each of the positions immediately following P which has not been marked with some number less than n must already have an option marked n. We continue in this way until it is impossible to mark any further node with any ordinal number, and then attach the symbol ∞ to any remaining nodes (which we call *unmarked*). The *value* of a position marked n is then n, while the value of an unmarked position is the symbol ∞ followed by the values of all marked options as subscripts.

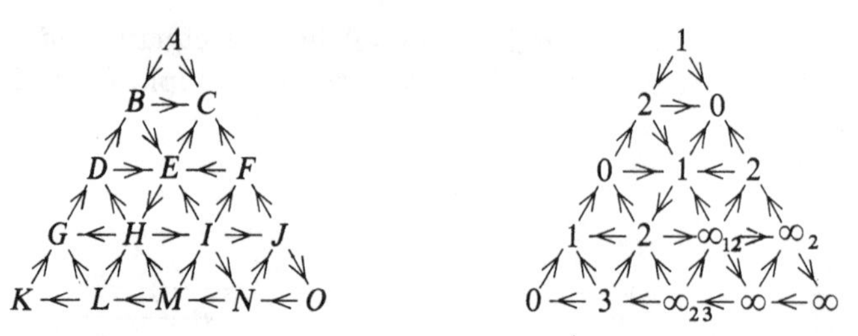

FIG. 31. A game which need not terminate, and its analysis.

Take for instance the graph of Fig. 31 suggested by Aviezri Fraenkel.

The reader will find it easier to understand the marking process if he draws the graph of Fig. 31, and follows our argument marking the various nodes in succession.

The node C has no exit move, and so we can label it 0. (If there had been no such node, then every node would be unmarked, and have value ∞.) Then D can be marked 0, because although B and E are as yet unmarked, they both have C as an option. Now G can be marked 1, since its only option is D, and K can be marked 0, since its only option is G.

At this point, the only node we can mark is E, which may and must be marked 1, since its only marked option is C (marked 0), and from its unmarked option H we can indeed get to G, marked 1. Now all Bs options are marked (with marks 0 and 1), so B is marked 2, and similarly A is marked 1, since its only options are marked 0 and 2, and F is marked 2.

Now the only node we can mark is H, which has options already marked 0 and 1, and an additional option I, from which we can get to F, already marked 2. So H is marked 2, and then L has only marked options, with marks 0, 1, 2, and so can be marked 3. With this, we obtain the marks given in the right hand part of the Figure.

The nodes I, J, M, N, O are unmarked, for since the adjacent nodes do not include a zero mark, the only plausible mark is 0. But each of I, J, M, N, O has another of I, J, M, N, O as an option, and from this option we cannot get to any node already marked 0. So we attach the symbol ∞ to each of these nodes, with subscripts as appropriate. Thus since from M we can move to H and L, with marks 2 and 3, we have written ∞_{23} for M.

Now we assert that in play, a position marked n behaves like the Nim-heap $*n$, and so in particular, is a second player win if and only if $n = 0$, and otherwise a first player win. Also, an unmarked position is a first player win if and only if it has some subscript 0, and is otherwise a draw.

We further assert that the disjunctive sum $P + Q$ of positions P and Q will be marked if and only if P and Q are marked, and that then its mark will be the Nim-sum of those of P and Q. We explain these assertions after considering an example.

TRAFFIC JAMS

Figure 31 may be considered as a map of a fictitious country, with towns whose names run from Aberystwyth to Oswestry, and one-way motorways between them. Four vehicles are placed initially at Aberystwyth, Dolgellau, Ffestiniog, and Merioneth, and either player, when it is his turn to move, may move any of these vehicles from one town to any adjacent one along a motorway (in the right direction). Each town is big enough to accommodate all four vehicles at once, should the need arise. When all the vehicles are stranded at Conway (from which there is no escape), the player about to move loses, for he is unable to do so. What should the first player do?

He should observe first that the game is a disjunctive sum of four smaller games, one for each vehicle, so that he should make some move to a sum of value 0, if possible. (Note that 0 is the only second player win, and that after making a move one is the second player.) Now the vehicles at A, D, F have finite marks, but that at M has value ∞, and so must be moved if the sum is to be marked, and to a town marked 3 if we are to move to a 0 position, since $3 = 0 +_2 1 +_2 2$. We conclude that the unique winning move here is the one in which the vehicle which was originally at Merioneth is moved to Llanfairpwllgwyngyllgogerychwyrndrobwllllantysiliogogogoch.

(How long would it take a professional Chess player to see this?)

Of course, *any* position with a vehicle at Novosibirsk is a draw.

To see that the theory works, we first observe that the positions we mark n are really Nim-heaps with reversible moves, for from such a position we can certainly get to positions with any desired marks less than n, and from all other options we can move to "simpler" positions marked n. (This is really a proof by induction on the order in which we mark the positions.)

So by the theory of generalised forms of Nim, the sum of two such positions with marks a and b is another, with mark $a +_2 b$. So the only assertion we need prove is that if $P + Q$ is marked, so are P and Q. Take an earliest marked position $P + Q$ for which this fails, and let a be the mex of the marks at options of P, and b the corresponding mex for Q. We shall show that in fact P is marked a and Q is marked b.

If P, say, were unmarked, then some option P' of P would also be unmarked. Then $P' + Q$ is unmarked (by induction), and so since $P + Q$ is marked, $P' + Q$ must have some option $P'' + Q$ or $P' + Q'$ which is marked with the same mark as $P + Q$. The latter case is impossible by induction since P' is unmarked, and so $P'' + Q$ must be marked, so that P'' and Q must both be marked, by induction. The mark of Q can only be b, and, since $P'' + Q$ and $P + Q$ have the same mark, the mark of P'' can only be a.

We have shown that every P' which is unmarked must have an option P'' marked a, and it follows that P should have been marked a, proving our assertion.

CHAPTER 12

How to Lose when you Must

(The misère theory of disjunctive sums)

In the pleasant orchard closes,
'God bless all our gains', say we;
But 'May God bless all our losses'
Better suits with our degree.
Elizabeth Barrett Browning, "The Lost Bower"

This chapter, and the two that follow it, are a digression from our main theme to consider various other generalisations of the theory of impartial games. The reader who does not wish to have his train of thought disturbed should pass at once to Chapter 15.

MISÈRE PLAY OF DISJUNCTIVE SUMS

We have seen that when the last player able to move is defined to be the winner (the *normal play* rule), the theory of disjunctive sums is really very simple. Each component behaves like a Nim-heap of a certain size (its Grundy number), and we can simply imagine ourselves playing Nim. It is remarkable that when we simply change the rules by declaring that the last player able to move is the loser (the *misère play* rule), the situation changes completely, and the whole theory becomes much more complicated. Nevertheless, there is a lot we can say, and in many cases we are able to give a complete analysis of some quite complicated games.

MISÈRE NIM

The strategy here is due to Bouton himself (indeed, if anything, misère Nim is more commonly played than normal Nim):

Play as you would in normal Nim, making the Nim-sum of the heap-sizes zero, unless your move would leave only heaps of size one, (discounting empty heaps). In this case, move so as to leave either one more or one fewer one-heaps than the normal play move.

In other words a position with some heap of size two or more is a second player win if and only if its Nim-sum is zero, but one in which all the heaps are zero or one is a second player win only if its Nim-sum is one (i.e. it has an *odd* number of one-heaps).

There is another way to describe this strategy which will be useful later—every Nim-position has a *Nim-type*, which is one of the symbols $0, 1, 0^0, 1^1, 2^2, 3^3, \ldots$. Given the Nim-types of two positions, we can determine the Nim-type of their sum by using the rules $0 + T = T$ (for any type), $1 + 1 = 0$, $1 + t^t = u^u$ $(u = 1 +_2 t)$, $a^a + b^b = c^c$ $(c = a +_2 b)$ which are perhaps easiest seen from the following table

	0	1	0^0	1^1	2^2	3^3	4^4	5^5	$\ldots$
0	0	1	0^0	1^1	2^2	3^3	4^4	5^5	
1	1	0	1^1	0^0	3^3	2^2	5^5	4^4	
0^0	0^0	1^1	0^0	1^1	2^2	3^3	4^4	5^5	
1^1	1^1	0^0	1^1	0^0	3^3	2^2	5^5	4^4	
2^2	2^2	3^3	2^2	3^3	0^0	1^1	6^6	7^7	
3^3	3^3	2^2	3^3	2^2	1^1	0^0	7^7	6^6	
4^4	4^4	5^5	4^4	5^5	6^6	7^7	0^0	1^1	
5^5	5^5	4^4	5^5	4^4	7^7	6^6	1^1	0^0	

In other words, the types a^a combine by ordinary Nim-addition, but there are two additional types 0 and 1. A single Nim-position has type n^n if its Nim-sum is n, *unless* all its heaps have sizes 0 or 1, when its type is 0 or 1 according as there are an even or odd number of 1's. Finally, the type of a position determines its outcome—in normal Nim the wins for the second player are 0 and 0^0 (and so we need not distinguish between these), but in misère Nim they are 1 and 0^0 (so that we must).

There are many other games for which the same system of types works, and many people have guessed that the theory of misère Nim is a prototype for that of misère sums in general. The prevailing belief seems to be that a good strategy is "play as in normal play until the game is nearly over, and then make a sensible move". But Grundy showed that in general the situation can be much more complicated than this allows.

REVERSIBLE MOVES

We use the notation of the previous chapter, so that $\{A, B, C, \ldots\} = G$ denotes a game G from which either player can move to any one of $A, B, C, \ldots$. Grundy discovered one way of simplifying games, which turns out to be the only way, namely by pruning reversible moves. We prefer to describe first the opposite notion, where we make a game more complicated (without any real change) by adding new reversible moves.

Let $G = \{A, B, C, \ldots\}$. Then we say that the game

$$H = \{A, B, C, \ldots, X, Y, Z, \ldots\}$$

has been obtained from G by *adding reversible moves* if there are moves from each of the new positions X, Y, Z, ... back to G, *provided that*, if G is empty, one of X, Y, Z, ... is a second player win.

The last clause is known as *the proviso*. Then Grundy's principle is:

The outcome of a sum of games is not affected by replacing G by H, (or vice versa), if H can be obtained from G by adding reversible moves, subject to the proviso.

For supposing we have a strategy for winning such a sum, with G being one of the summands. Play "the same" strategy when G is replaced by H, never yourself making use of the new moves. If your opponent does so, moving from H to X, say, then you should "reverse" the position to G, *provided* H is not the only non-zero component remaining, when you should instead move to the second player win position which is accessible from H.

Observe that, with the exception corresponding to the proviso, this is the same argument as in normal play. The extra complication arises because of the unnatural treatment of 0, which is now counted as a win for the first player even though he has no good move. Unfortunately, the complications so produced persist indefinitely, and make the misère play theory much more complicated than the normal one.

Suppose $H = \{A, B, C, \ldots, X, Y, Z, \ldots\}$ has been obtained by adding reversible options to $G = \{A, B, C, \ldots\}$. Then when H occurs in some sum we should naturally like to replace it by the simpler game G. Of course we will normally be given only H, and have to find the simpler game G for ourselves. How do we do this?

Here are two observations which make this fairly easy:

(i) G must be obtained by deleting certain options of H.

(ii) G must itself be an option of any of the deleted options of H, and so G must itself be a *second option* of H, if we can delete any option at all.

On the other hand, if we obey (i) and (ii), the deletion is permissible, except that we can only delete *all* the options of H (making $G = 0$) if one of them is a second player win.

It turns out that if we make all possible such deletions at all positions of some game G, we obtain the unique simplest possible form of G—in other words, no further simplifications are possible. We shall prove this later. It was formerly known to some people as *Grundy's conjecture*, although Professor Smith informs me that in fact Grundy conjectured no such thing, and firmly believed the opposite!

THE BEHAVIOUR OF NIM-HEAPS

We first examine games whose options are all Nim-heaps. As usual, these are defined by

$$0 = \{\}, 1 = \{0\}, 2 = \{0, 1\}, \ldots, n = \{0, 1, \ldots, n - 1\},$$

and indeed for all ordinals α by $\alpha = \{\beta < \alpha\}$. In this chapter, we shall omit the prefixed stars.

THEOREM 72. *A game G whose options are all Nim-heaps reduces to a Nim-heap itself unless all the options have size at least 2. When it reduces, it is to the least Nim-heap not appearing as an option.*

Proof. If the game reduces, it must be to some second option of itself, and so to some Nim-heap, which must obviously be the least Nim-heap not an option. If this is not 0, the reduction does in fact take place, and if it *is* 0, it will still take place if some option was 1 (a second player win).

So for example, we have $\{1\} = 0$, $\{0, 1, 3, 5\} = 2$, $\{1, 2, 3\} = 0$, but $\{2\}$ does not reduce to a Nim-heap, at least by our rule. In fact $\{2\}$ is a second player win, and so would have to reduce to 1, if to any Nim-heap, but this is impossible, since $\{2\} + \{2\}$ is also a second player win, whereas $1 + 1$ is not.

A fairly immediate corollary of Theorem 72, which we state without proof, is:

If n is a Nim-heap, so is n + 1. Its size is given by the normal Nim-addition rule.

Thus $2 + 1 = 3$, $3 + 1 = 2$, $1 + 1 = 0$, etc. On the other hand, the game $2 + 2 = \{0 + 2, 1 + 2, 2 + 0, 2 + 1\} = \{2, 3\}$ (by this rule), which does not reduce to a Nim-heap.

We postpone formal proof that the Grundy principle gives all possible simplifications until later, and use it now to discuss simplest forms of the smallest games. If we made no simplification at all, we should find 1 game "born on day 0", 2 games born by day 1, 4 games born by day 2, $2^4 = 16$ by day 3, $2^{16} = 65536$ by day 4, 2^{65536} games born by day 5, and so on, since any subset of the games born by day n yields a game born by day $n + 1$ (at most).

When we prune reversible moves, we get slightly smaller numbers. Counting only games in simplest form, Grundy and Smith showed that there was

1 game born by day 0, 2 by day 1, 3 by day 2, 5 by day 3, 22 by day 4, and 4171780 by day 5. We extend their list one place by remarking that there are exactly

$$2^{4171780} - 2^{2095104} - 3{\cdot}2^{2094593} - 2^{2094081} - 3{\cdot}2^{2091522} - 2^{2088960}$$
$$-3{\cdot}2^{2088448} - 2^{2087937} - 2^{2086912} - 2^{2086657} - 2^{2086401} - 2^{2086145}$$
$$-2^{2085888} - 2^{2079234} + 2^{1960962} + 21$$

games in simplest form born by day 6. (This number is more than

$$99{\cdot}9999999999999999999999999999999999\,\%$$

of $2^{4171780}$. The number of games in simplest form born by day 7 is very small compared to $2^{2^{4171780}}$, but huge compared to $2^{2^{4171779}}$.)

It is not hard to show that for a suitable real number γ_0 (approximately $0{\cdot}149027998351785\ldots$) if we define $\gamma_{n+1} = 2^{\gamma_n}$ then the number of games in simplest form born by day n is the next integer above γ_n.

The day on which a game is born tells us how long it can possibly last (if it is less than ω), so we call it the *length* of the game. On the next page we draw trees for the 22 reduced games of length at most 4. Since an abbreviated notation rapidly becomes almost essential, we use $ABC\ldots$ for $\{A, B, C, \ldots\}$, except that we use A_+ for $\{A\}$ to distinguish this from A itself. The 22 games of length at most 4 are

$$0,\ 1,\ 2,\ 3,\ 4,\ 2_+,\ 3_+,\ 32$$
$$2_{++},\ 2_+0,\ 2_+1,\ 2_+2,\ 2_+20,\ 2_+21,\ 2_+210,$$
$$2_+3,\ 2_+30,\ 2_+31,\ 2_+32,\ 2_+320,\ 2_+321,\ 2_+3210.$$

THE MISÈRE GRUNDY VALUE

The normal Grundy number of G can be defined as the unique number n for which the disjunctive sum $G + n$ is a second player win in normal play. This number we shall call in this chapter $\mathcal{G}^+(G)$. Similarly, we can define the misère Grundy number $\mathcal{G}^-(G)$ to be the unique n for which $G + n$ is a second player win in misère play. It can easily be calculated, and hence shown to exist, by the rule:

> $\mathcal{G}^-(0) = 1$. Otherwise, $\mathcal{G}^-(G)$ is the least number (from 0, 1, 2, ...) which is not the $\mathcal{G}^-$-value of any option of G.

Notice that this is just like the ordinary "mex" rule for computing $\mathcal{G}^+$, except that we have $\mathcal{G}^-(0) = 1$, $\mathcal{G}^+(0) = 0$.

In the analysis of many games, we need even more information than is

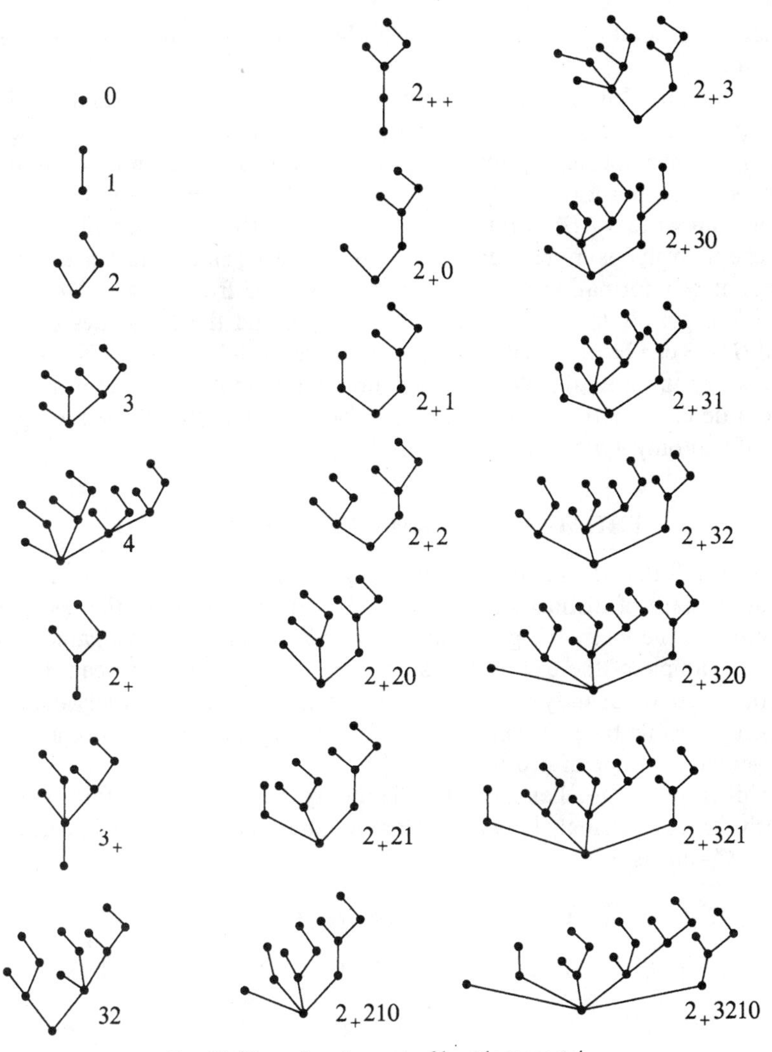

FIG. 32. The reduced games of length at most 4.

provided by either of these values, and so we shall define a more complicated symbol that we call the *$\mathscr{G}$*-value*, $\mathscr{G}*(G)$. This is the symbol

$$g^{g_0 g_1 g_2 \cdots}, \text{ where} \begin{cases} g = \mathscr{G}^+(G) \\ g_0 = \mathscr{G}^-(G) \\ g_1 = \mathscr{G}^-(G+2) \\ g_2 = \mathscr{G}^-(G+2+2) \\ \quad \cdots \end{cases}$$

where in general g_n is the $\mathscr{G}^-$-value of the sum of G with n other games all equal to 2.

We already know how to compute the leading entries g and g_0. Each remaining entry g_{n+1} is the least number not equal to g_n or $g_n +_2 1$, and not the g_{n+1}-entry for any option of G. The $\mathscr{G}*$-value is apparently an infinitely long symbol, but fortunately $g_{n+1} = g_n +_2 2$ for all sufficiently large n, and so we can write $\mathscr{G}*(G) = g^{g_0 g_1 \cdots g_m}$ to mean that this holds for all $n \geqslant m$.

The $\mathscr{G}*$-value is more useful than it might seem, since a rather surprising amount of information about G can be recovered from it. The $\mathscr{G}*$-value of $G + 2$ is simply $h^{g_1 g_2 g_3 \cdots}$, where $h = g +_2 2$, and the $\mathscr{G}*$-values of $G + 1$ and $G + 3$ can be obtained from those of G and $G + 2$ by simply Nim-adding 1 to every entry. So the $\mathscr{G}*$-value of G determines that of $G + n$ for all $n < 4$, and it determines the *outcome* of $G + n$ for all n, since this is a second player win if and only if $g_0 = n$.

THE MISÈRE FORM OF GRUNDY'S GAME

We recall that the move in Grundy's game is to divide any heap into two smaller heaps of distinct sizes. Of course in the misère form the last player to move is the loser. We give a fairly extended analysis of this game, partly as an example of the use of $\mathscr{G}*$-values, and partly because we can disprove a conjecture of Grundy's that the second player wins are precisely the heaps of size divisible by 3. It will turn out that a heap of size 48 is not a second player win, but one of size 50 is.

This is quite a good example for the theory, because the positions simplify nicely for a surprisingly long time. Here is a short list of simplified forms, with their $\mathscr{G}^*$-values:

$n =$	0	1	2	3	4	5	6	7	8	9	10	11	12	13	14	15	16	17
$G_n =$	0	0	0	1	0	2	1	0	2	1	0	2	1	3^{143a}	2	1	3^{143a}	2

$n =$	18	19	20	21	22	...
$G_n =$	4^{056b}	3^{143a}	0^{2c}	4^{056b}	3^{143d}	...

$$a = 2_2 21,\ b = a_2 a20,\ c = ba3,\ d = cba_1 2_2 1.$$

When the reduced form is a Nim-heap, of size n, we have simply written n. Otherwise, we give the complete $\mathscr{G}*$-value, followed by a small letter which is the name of the game, while below the list the structure of this game is described in more detail. In the abbreviated names for these games, a_n denotes the reduced form of $a + n$, so for instance 2_2 denotes the reduced form of $2 + 2$, namely $\{3, 2\}$, or just 32 in the abbreviated notation. The reader will see why we are at such pains to abbreviate the notation if he examines the tree for the game b given as Fig. 33!

To see how the table was computed, we take $n = 16$ and $n = 17$. In the first case, we have at first sight 7 options

$$G_1 + G_{15}, G_2 + G_{14}, \ldots, G_7 + G_9.$$

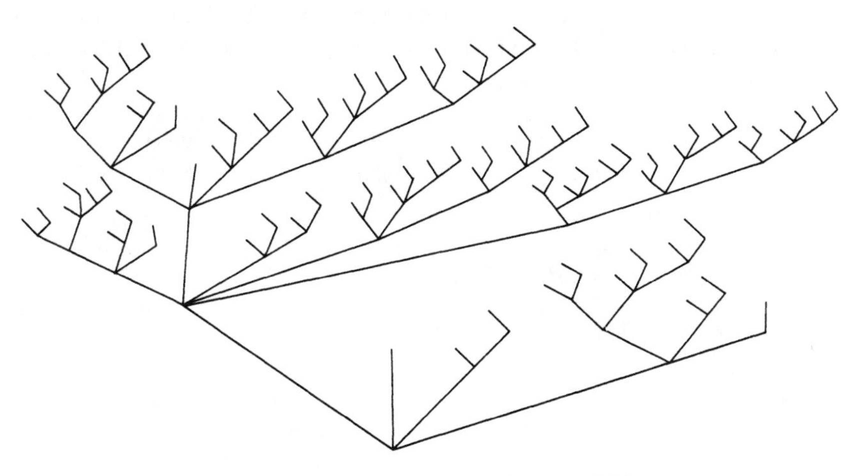

FIG. 33. The game $b = a_2 a\, 20$, where $a = 2_2 21$.

Making use of the previous entries, these become the 7 games

$$0 + 1, 0 + 2, 1 + a, 0 + 1, 2 + 2, 1 + 0, 0 + 1,$$

which simplify to $1, 2, a_1, 1, 2_2, 1, 1$, so that $G_{16} = \{a_1, 2_2, 2, 1\}$. Here we can delete the option a_1 to obtain the reduced form $\{2_2, 2, 1\} = a$, since a_1 has this as an option. In a similar way, G_{17} has

$$0 + a, 0 + 1, 1 + 2, 0 + a, 2 + 1, 1 + 2, 0 + 0, 2 + 1$$

as options, which simplify (using $2 + 1 = 3$) to $a, 1, 3, a, 3, 3, 0, 3$, so that $G_{17} = \{a, 3, 1, 0\}$, which we can simplify to $\{1, 0\} = 2$, since both a and 3 have 2 as an option.

The game $c = ba3$ has options with $\mathcal{G}*$-values 4^{056}, 3^{143}, 3^3, and we compute its $\mathcal{G}*$-value $0^{202020\cdots}$ from the following "sum":

$$\begin{array}{l} 4^{056464\ldots} \\ 3^{143131\ldots} \\ 3^{313131\ldots} \\ \hline 0^{202020\ldots} \end{array}$$

in which each number is the least number not in any of the corresponding places immediately above it, and which is not the previous superscript, or that superscript Nim-summed with 1.

So the table is fairly easy to compute, and from the known properties of the $\mathscr{G}*$-values, we see that it gives the outcome of any position in Grundy's game in which all heap-sizes are at most 22, and at most one is 13, 16, or 18 or more.

The table in Fig. 34 extends this discussion to $n = 50$. The first entry for a given n is the $\mathscr{G}*$-value of G_n, and later entries give the $\mathscr{G}*$-values of the sums $G + a$, $G + b$, $G + c$, ..., etc.

$1:0$	$13:3^{143a}$	$25:3^{143}$	$37:1^1 \mid 2^2$	$49:1^1$
$2:0$	$14:2$	$26:0^2$	$38:2^{47}$	$50:0^{043}$
$3:1$	$15:1$	$27:4^{056}$	$39:4^{056}$	
$4:0$	$16:3^{143a}$	$28:1^1 \mid 2^2 \mid 5^{47} \mid 1^1 \mid 2^2$	$40:1^1$	$a = 2_2 21$
$5:2$	$17:2$	$29:2^2 \mid 1^{43} \mid 6^{56} \mid 2^0$	$41:5^{47}$	$b = a_2 a20$
$6:1$	$18:4^{056b}$	$30:3^0 \mid 0^3 \mid 7^6 \mid 3^2$	$42:4^{056}$	$c = ba3$
$7:0$	$19:3^{143a}$	$31:1^1 \mid 2^2 \mid 5^{47}$	$43:1^1$	$d = cab_1 2_2 1$
$8:2$	$20:0^{2c}$	$32:2^2 \mid 1^{43} \mid 6^{16}$	$44:5^{47}$	$e = dc_1 b_2 ba3$
$9:1$	$21:4^{056b}$	$33:4^0 \mid 7^3$	$45:4^{052}$	$f = edb_1 ca_2 20.$
$10:0$	$22:3^{143d}$	$34:1^1 \mid 2^2$	$46:1^1$	
$11:2$	$23:0^{2e}$	$35:2^{47} \mid 1^{096}$	$47:5^{47}$	
$12:1$	$24:4^{056f}$	$36:4^0 \mid 7^3$	$48:4^2$	

$\mathscr{G}*$-values of sums of a, b, c:

	0	a	$a+a$	0	a	$a+a$	0	a	$a+a$
0	0^{12}	3^{143}	0^{12}	0^2	3^{43}	0^2	0^0	3^3	0^0
b	4^{056}	7^{0587}	4^{056}	4^{56}	7^{587}	4^{56}	4^4	7^7	4^4
$b+b$	0^{12}	3^{143}	0^{12}	0^2	3^{43}	0^2	0^0	3^3	0^0
	a's and *b*'s only			with one c			two or more *c*'s.		

FIG. 34. A table for Grundy's game with misère play.

In the table for sums of a, b, c the values given suffice to show that each layer has period 2 in both directions—in other words that in such sums we can treat $a + a$ and $b + b$ as 0. But for c, the corresponding pretence is $c + c + c = c + c$.

It can actually be shown, although this is not needed for the analysis of the 50-heap, that values G_n for $22 \leqslant n \leqslant 27$ can be equated with the appropriate one of a, b, c provided there is no larger heap. The table is complete for the posterity of a 50-heap, and since the outcome depends on the equation $g_0 = 0$, we see that 48 is *not* a second player win, and that 50 *is*, disproving the aforementioned conjecture of Grundy in two ways.

It would become intolerably tedious to push this sort of analysis much further, and I think there is no practicable way of finding the outcome of G_n for much larger n.

A similar analysis is given for three other games in Fig. 35. For many octal games a complete analysis is possible because the reduced forms are all Nim-heaps, and in others because the reduced forms are all *tame* in the sense defined below. This is true, for instance, of the games PRIM and DIM of the last chapter. Although it is not true for Kayles, a more subtle complete analysis of that game has now been given by Conway and Sibert.

n	Kayles	·4	·6
0	0	0	0
1	1	0	0
2	2	0	1
3	3	1	2
4	1	1	0
5	$4^{14\alpha}$	2	1
6	3	0	2
7	$2^{2\beta}$	3	3^{143a}
8	3_2	1	1
9	$4^{04\gamma}$	1	2
10		0	3^{143a}
11		3^{143a}	4^{056b}
12		3	0^{2c}
13		2^{052d}	
14		2	

Recall that:
Kayles is the game ·77 in which one can remove either 1 or 2 adjacent counters from a line.
In its analysis we have $\alpha = 2_2 321$, $\beta = \alpha_1 \alpha 3_2 2_2 30$, $\gamma = \beta_1 \beta \alpha_3 \alpha_2 3_2 20$.
In ·4 we can remove a single counter which is not at the end of any row, and in ·6, any single counter which is not isolated (i.e. not the whole of its row). The games a, b, c arising in their analysis are $a = 2_2 21$, $b = a_2 a320$, $c = ba3$. The game β behaves like 2, and b behaves like a_1. The game d *is* $a3_2 0$

FIG. 35. Misère tables for three simple games.

TAME GAMES, AND RESTIVE AND RESTLESS GAMES

The only $\mathcal{G}*$-values which arise for Nim-positions are 0^{12}, 1^{03}, and n^n ($n = 0, 1, 2, \ldots$), which correspond to what we earlier called the *Nim-types* $0, 1, n^n$. If all the positions (including the initial position) of some game G have $\mathcal{G}*$-values selected from this population 0^{12}, 1^{03}, n^n, we call G *tame*. Then:

THEOREM 73. *The sum of tame games is tame. To compute its* $\mathcal{G}*$*-value, we can replace the summands by Nim-positions of the same* $\mathcal{G}*$*-values, and take the* $\mathcal{G}*$*-value of the resulting sum.*

We do not bother to prove this in detail, remarking merely that it follows fairly easily by generalising the strategy for misère Nim.

It is handy to indicate that a game is tame by putting $*$ at the end of its

$\mathcal{G}*$-value. Then we can further abbreviate the values 0^{12*} and 1^{03*} to just $0*$ and $1*$. So we write the $\mathcal{G}*$-values of tame games just as $0*$, $1*$, or n^n*. With this additional convention, and the convention of writing the $\mathcal{G}*$-value of n itself as n, our abbreviations become even more useful, for if we know the $\mathcal{G}*$-values of tame games G and H, we can compute that of $G + H$.

If all the options of G are tame, but G is not, then we call G *restive* or *restless.* For such games, with $\mathcal{G}*$-value $g^{g_0 g_1 g_2 \cdots}$, then *just one* of g, g_0 is 0 or 1, the other being 2 or more. Since the behaviour is quite different in the two cases, we call G *restive* if $g = 0$ or 1, and *restless* if $g_0 = 0$ or 1.

From the $\mathcal{G}*$-value of a *restive* G, we can work out the outcome of the sum of G with any Nim-position N, and indeed the $\mathcal{G}*$-value of any such compound. The rule is:

Mentally replace G by a Nim-heap of size g_0, *if all the heaps in N have sizes* 0, 1, g_0, *or* $g_0 +_2 1$, *and of size g if not. Then the misère outcome of G is the same as the normal outcome of the resulting Nim-position.*

If the heaps of N have Nim-sum n, then the $\mathcal{G}*$-value of $G + N$ is found from that of G by Nim-adding n to every entry, if each heap-size is $0, 1, g_0$, or $g_0 +_2 1$, and is otherwise m^m, where $m = g +_2 n$.

On the other hand, there is no easy rule for finding the outcome of $G + N$ for a *restless* G and an arbitrary Nim-position N. The tables show just how badly such sums can behave. We have chosen here the particular tables likely to be useful for simple games.

Note that in a sense, restive games are *ambivalent Nim-heaps,* which choose their size (g_0 or g) according to the company. There are many other games which exhibit behaviour of this type, and it would be very interesting to have some general theory for them.

SOME TABLES FOR RESTLESS GAMES

We tabulate $\mathcal{G}^-(G + a + b)$, where a and b are the row and column headings. This is the value of c such that $G + a + b + c$ is a P-position. Since Nim-addition of 1 to a or b produces Nim-addition of 1 to c we only tabulate c for even a and b.

	$b =$ 0	2	4	6	8	10	12	14	16	18	20	22	...
0	1	4	2	8	6	12	10	16	14	20	18	24	...
2	4	2	0	6	8	10	12	14	16	18	20	22	...
$a =$ 4	2	0	4	10	14	6	16	8	12	24	22	20	...
6	8	6	10	2	0	4	14	12	18	16	24	26	...
8	6	8	14	0	2	14	18	4	10	...			

Table for many restless games with $\mathcal{G}* = 2^{142}$ (examples 2_21, 2_231)

	$b = 0$	2	4	6	8	10	12	14	16	18	20	22	...
0	1	4	2	8	6	12	10	16	14	20	18	24	...
2	4	3	0	7	9	11	13	15	17	19	21	23	
$a = 4$	2	0	4	10	15	6	12	9	18	...			
6	8	7	10	3	0	4	14	12	20	...			
8	6	9	15	0	3	17	...						

Table for some restless games with $\mathscr{G}* = 3^{143}$ (example $2_2 21$)

	$b = 0$	2	4	6	8	10	12	14	16	18	20	22	...
0	1	4	2	8	6	12	10	16	14	20	18	24	...
2	4	6	0	2	12	14	8	10	20	22	16	18	...
$a = 4$	2	0	4	6	8	10	12	14	16	18	20	22	...
6	8	2	6	4	0	16	14	12	10	24	22	20	...
8	6	12	8	0	4	18	2	20	22	10	14	16	...

Table for many restless games with $\mathscr{G}* = 4^{14}$ (example $2_2 321$)

In each of these cases we can substitute for the options any other tame games with the same values of $\mathscr{G}*$, so that for instance $2_+ 2_{22} 1$ has the second of the three tables, since 2_+ is a tame game with $\mathscr{G}* = 0^{0202\ldots} = \mathscr{G}*(2_2)$, and 2_{22} $(=2 + 2 + 2)$ is a tame game with $\mathscr{G}* = 2^{2020\ldots} = \mathscr{G}*(2)$.

The simplest restless games with $\mathscr{G}* = 3^{053}$, 2^{052}, and 5^{05} have tables obtained by Nim-adding 1 to every entry in these three tables. Also, the games $3_2 0$ and $3_2 20$ have a table obtained from the first one above by replacing the leading entry by 0, and the games $3_2 20$ and $3_2 320$ have tables obtained similarly from the second and third tables above. We can replace the options in these cases also by other tame games with the same $\mathscr{G}*$-values.

We now show that Grundy's method of pruning reversible moves gives all possible simplifications.

Definitions. The *misère outcome* $o^-(G)$ is the symbol P or N according as the Previous or Next player to move has a winning strategy, in misère play.

G is *like* H iff $o^-(G + T) = o^-(H + T)$ for all T

G is *linked* to H (by T) iff $o^-(G + T) = o^-(H + T) = P$ for some T.

When G and H are unlike, any game T for which $G + T$ and $H + T$ have distinct outcomes will be said to *distinguish* between G and H. Finally, for any game G we define its *mate*, G^-, to be the game obtained from G by interchanging 0 and 1 whenever they arise as positions of G—so $0^- = 1$, $1^- = 0$, and otherwise $\{A, B, C, \ldots\}^- = \{A^-, B^-, C^-, \ldots\}$. So that we can point out certain analogies, we ask the reader to define the normal outcome $o^+(G)$, and to define $G^+ = G$ for all G. It is immediate from the definitions that the misère outcome of G is the normal outcome of G^-, and vice versa.

THEOREM 74. *The misère outcome of* $G + G^-$ *is always* P.
(Compare: the normal outcome of $G + G$ is always P.)

Proof. When your opponent moves to a position H in one component, reply by moving to H^- in the other, until eventually the game reduces to $0 + 1$ or $1 + 0$, a P position. (In the normal play analogue, we eventually get to the P-position $0 + 0$.)

LEMMA. *If* G *and* H *are unlike, there is a game* T *with* $G + T$ *a* P *position,* $H + T$ *an* N *position.*

Proof. If not, there will certainly be some U with $G + U$ an N position and $H + U$ a P position. Then take $T = \{A^-, B^-, C^-, \ldots, U\}$ when $G = \{A, B, C, \ldots\}$, and observe:

$H + T$ is N, since the next player can move to $H + U$.

$G + T$ is P, since the options $A + T, \ldots, G + A^-, \ldots, G + U$

are all N. ($A + T$ and $G + A^-$ have the P option $A + A^-$.)

Now for the first of our main results.

THEOREM 75. *G is like H if and only if*
(i) *G is linked to no option of H,*
(ii) *H is linked to no option of G, and*
(iii) *G and H have the same outcome if either is* 0.

Proof. For if T links G to some option of H, or H to some option of G, then T distinguishes between G and H, and if G and H have distinct outcomes, 0 distinguishes between G and H. So for G to be like H, (i), (ii), and (iii) must hold. Supposing that they do all hold, we let T be a game for which $G + T$ is N, and prove that $H + T$ is also N, supposing that $G + U$ and $H + U$ have the same outcome for all options U of T. It is plain that this suffices.

If $G + T$ is N, we have one of:

(a) $G = T = 0$
(b) $G' + T$ is P for some option G' of G
(c) $G + T'$ is P for some option T' of T.

In case (a), $G + T$ and $H + T$ both have outcome N, by (iii), since these games are just G and H.

In case (b), $H + T$ is N, since otherwise T would link G' to H.

In case (c), $H + T'$ is also P, by hypothesis, so that $H + T$ is N.

Now for our second main result:

THEOREM 76. *G is linked to H if and only if:*
(i) *G is like no option of H*
(ii) *H is like no option of G.*

Proof. For if T links G to H, then T distinguishes G from any option of H, and H from any option of G, so that (i) and (ii) must hold. Supposing that they do hold, then we can find for any option G^a of G a game U^a with $G^a + U^a$ a P position, $H + U^a$ an N position, and similarly for any option H^b of H a game V^b with $G + V^b$ an N position, but $H^b + V^b$ a P position. We let T be the game $\{U^a, \ldots, V^b, \ldots\}$ whose options are all these games $U^a, \ldots, V^b$ corresponding to all the options of G and H.

Then $G + T$ has options $G^a + T$, $G + U^a$, $G + V^b$, typically, of which the first two are N since they have the option $G^a + U^a$, which is P, and the third is N by supposition. Similarly, all options of $H + T$ are N, and so $G + T$ and $H + T$ will both be P, unless $G = H = 0$. But in the excepted case, 1 links G to H.

(In the normal play analogues of Theorem 75, the condition (iii) is absent, so that the analogue of Theorems 75 and 76 have the same form, and we inductively see that G is like H if and only if G is linked to H. We can inductively deduce from these theorems that G is like H if and only if G is like no option of H and H is like no option of G, and thence, inductively again, that every game is like some Nim-heap.)

THEOREM 77. *Suppose that neither G nor H has a reversible option, and that G is like H. Then every option of G is like some option of H, and every option of H like some option of G.*

Proof. Since G is not linked to H^b, we must have either some G^a like H^b, *or* G like some option of H^b. But in the second case, H^b would be a reversible option of H, and since the proviso is clearly satisfied, this contradicts our assumption. So every option H^b of H is like some option G^a of G, and vice versa.

THE SIMPLEST FORM OF A GAME FOR MISÈRE PLAY

We obtain the simplest form of a game G by making all possible simplifications of the following type. At any position H of G, we may delete certain options of H to obtain a simpler position K if and only if K is an option of each deleted option of H, and if K is zero, H was an N-position.

Theorem 77 plainly implies that if two like games G and H are both in simplest form, they are identical. So indeed Grundy's method of pruning reversible moves gives all possible simplifications, and we have proved all assertions made at the start of the chapter.

FURTHER DEVELOPMENTS

Since we are not interested in the distinctions that can be made between like games, we shall suppose from now on that all games are initially presented in simplest form. In the rest of the chapter we describe some theoretical results about the behaviour of games under addition. Since our results do not seem to have much application to practical game-playing, we do not give the proofs, which are surprisingly subtle in some cases.

SUBTRACTION OF GAMES

If $A + B = G$, we call A and B *parts* of G. It is natural in this case to write $A = G - B$. The *cancellation theorem* asserts that in fact A is determined by G and B. This theorem asserts that

(i) If $G + T$ and $H + T$ are like, so are G and H, and vice versa
(ii) If $G + T$ and $H + T$ are linked, so are G and H, and vice versa
(iii) T has only finitely many parts.

It seems curious that the induction requires all three parts.

Given the theorem, differences $G - H$ are unique when they exist, and in fact whenever $G - H$ exists, it equals $\{G' - H, G - H'\}$, where in the brackets any differences that do not exist are neglected. So we have an algorithm for subtraction—compute the game $\{G' - H, G - H'\}$, and then add it to H to see whether the sum is G. Of course, since the cancellation theorem tells us that games form an abelian semigroup with cancellation, we could in any case *adjoin* formal differences to obtain an abelian group, but I have not yet met any theorem whose proof can be simplified in this way.

EXTRAVERTED, INTROVERTED, AND DIVINE GAMES

Certain games exhibit quite surprising splittings into parts, as we shall see later. and in the study of this phenomenon the following notions seem to be useful. We call G *extraverted* if it has each of its options as a part, and *introverted* if it is a part of each of its options. We call G *divine* if whenever it is a part of every option of some game H, it is a part of H.

The *extraversion–introversion theorem* then asserts that G is extraverted if and only if it is divine, and if and only if it is a part of the game $G_+ = \{G\}$. Also, if G and H are extraverted, so are G_+, $G_* = \{G, G + 1\}$, and $G + H$, and the class of all extraverted games is precisely the closure of 0 under these operations. Finally, the only introverted games are 0 and 1, which are also the only games with negatives.

In particular, 2 is an extraverted game, and so is a part of 2_+. Using the subtraction algorithm, we find $2_+ - 2 = \{2_+, 2 - 1, 2 - 2\}$ which simplifies to $\{2_+3, 2_+, 0\}$. We draw the resulting equation in tree form:

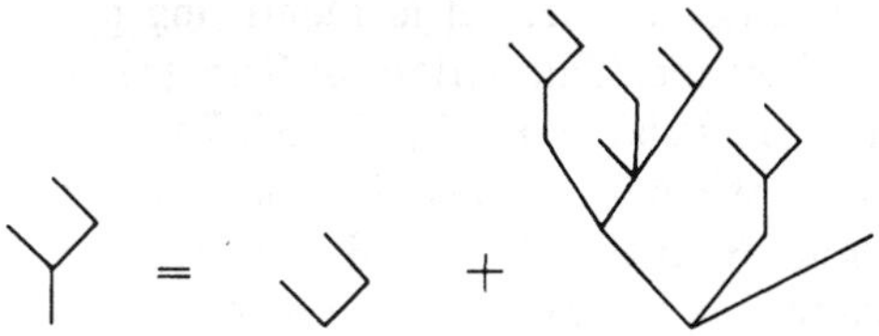

$$2_+ \quad = \quad 2 \quad + \quad 2_{+1}2_+0$$

Notice the rather remarkable fact that the whole is here simpler than one of the parts. I do not know of any game in which every part in every proper partition is more complicated than the whole.

EVEN, ODD, AND PRIME GAMES

We call the game G *even* when G is simpler than $G + 1$—more formally, when the simplest form of G is a position in the simplest form of $G + 1$. It can be shown that every G is either even or odd but not both, where H is called *odd* if it has the form $G + 1$, G even. Also, the sum of two even games is even, so that the even games form a subsemigroup of index two in the additive semigroup of all games.

We call the game P *prime* if in any partition $P = A + B$ either A or B is 0 or 1. These games are analogous to prime numbers in the multiplication of ordinary integers, and I conjectured at first that the partition of G into primes was unique. Note that the full form of the cancellation theorem shows that no game can have more than finitely many partitions into primes (neglecting parts 0 or 1), and we need only consider partition of even games into even primes. It can be shown that extraverted games do indeed have unique prime partitions, so that for instance the above partition of 2_+ is its only partition into even primes.

However, the following example, found jointly with Simon Norton, shows that certain games have more than one partition into primes. Let $G = (4+2)_+$. Then it can be shown that $G = 2 + K = 4 + L$, where $K = \{G, G + 1, 4\}$, $L = \{G, G + 1, K, K + 1, 2\}$. The fact that $G - 2$ and $G - 4$ exist and have these values follows from a slight strengthening of our remarks about subtraction—if all differences $G' - H$ and $G - H'$ exist (G', H' denote typical options of G and H), then so does $G - H$, unless perhaps when $G = 0$. Using this we can generalise the example so as to produce a game with an arbitrary finite number of distinct prime partitions. Further properties of the additive semigroup of games seem quite hard to establish—if $G + G = H + H$ is G necessarily equal to H or $H + 1$? If not, the group of game-differences has some non-trivial element of order 2.

The following remarks are helpful in identifying primes. If 0 or 1 is an option of G, then G is prime. If all options of G are prime, then so is G, *unless* G is one of the three particular games 2_+, 3_+, 32. These are extraverted games and so have unique prime partitions. That for 2_+ has been given above, and 3_+ has the similar partition $2 + 3_{+1}3_+1$, while $32 = 2 + 2$.

We end our comments on partitions with tree-diagrams for the unique prime partition of the game $(32)_+$ (Fig. 36):

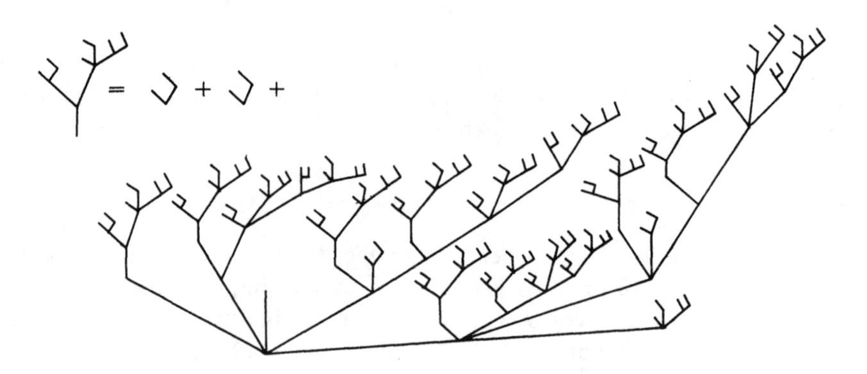

FIG. 36. The prime partition of $(32)_+$.

As a footnote to this chapter we tabulate the first few of the numbers γ_n defined on page 140. We have

$$\gamma_n = 2^{2^{2^{\cdot^{\cdot^{\cdot^{\gamma_0}}}}}}\Big\}n$$

where
$$\left.\begin{aligned}\gamma_0 &= 0{\cdot}149027\ldots\\ \gamma_1 &= 1{\cdot}108821\ldots\\ \gamma_2 &= 2{\cdot}156694\ldots\\ \gamma_3 &= 4{\cdot}458922\ldots\\ \gamma_4 &= 21{\cdot}992232\ldots\\ \gamma_5 &= 4171779{\cdot}999999999\ldots\end{aligned}\right\}$$

CHAPTER 13

Animating Functions, Welter's Game, and Hackenbush Unrestrained

Fallen from his high estate,
And welt'ring in his blood.
John Dryden, "Alexander's Feast"

There are some impartial games whose theory depends on unexpected interrelations between ordinary addition and Nim-addition.

WELTER'S GAME

This game is played on a semi-infinite strip with a finite number of coins, at most one per square. The squares are numbered with the non-negative integers 0, 1, 2, 3, . . . from the left end of the strip, as in Fig. 37. The legal move (for either player) is to move any one coin from its present square to *any*

0	1	⊘	⊘	4	⊘	6	⊘	8	9	10	⊘	12	⊘	14	15	16	⊘	18	⊘	20	• • •

FIG. 37. A position in Welter's game.

unoccupied square with a lower number. Thus, like the Silver Dollar game, Welter's game ends when one player (the loser) is unable to move because the coins are jammed in the lowest possible numbered squares 0, 1, 2, . . . , k. Welter's game differs from the Silver Dollar game in that any coin is allowed to bypass others in the course of a move.

We shall write $[a \mid b \mid c \mid \ldots]$ for the Grundy number of the position in this game when the coins are on the squares numbered $a, b, c, \ldots$. It is easy to check that $[a] = a$, and that $[a \mid b] = (a +_2 b) - 1$, so that both nim-addition and ordinary addition are involved in the theory of Welter's game. The full theory is surprisingly complex, and we shall be able to give it only after a detailed analysis of functions involving both kinds of addition, but for the benefit of the reader who does not wish to follow the detailed argument, we first give a computation rule for $[a \mid b \mid c \mid \ldots \mid k]$.

We take as our example the case $[2 \mid 3 \mid 5 \mid 7 \mid 11 \mid 13 \mid 17 \mid 19 \mid 23]$. We examine the numbers to see which pair are congruent to each other modulo the highest possible power of 2, and then we take any such pair out as *mates*. In our example, we have the congruences $3 \equiv 19$ and $7 \equiv 23 \pmod{16}$ but no congruences (mod 32), so we can take either (3, 19) or (7, 23) as our best mated pair.

Having removed the best mated pair, we treat the remaining numbers in the same way, obtaining a succession of pairs of *mates*, with at most one number (the *spinster*) left unmated. In our example, the mates are (3, 19), (7, 23), (5, 13), (11, 17), and there is a spinster 2. Then if $(a, b), (c, d), \ldots$ are the mates, we have the equation

$$[a \mid b \mid c \mid d \mid \ldots] = [a \mid b] +_2 [c \mid d] +_2 \ldots$$

if there is no spinster, and

$$[a \mid b \mid c \mid d \mid \ldots] = [a \mid b] +_2 [c \mid d] +_2 \ldots +_2 [s]$$

if there is a spinster s. The value is then computed using the formulae $[a \mid b] = (a +_2 b) - 1$, $[s] = s$, which we have already noted.

So we have the number

$$\begin{aligned} &[3 \mid 19] +_2 [7 \mid 23] +_2 [5 \mid 13] +_2 [11 \mid 17] +_2 [2] \\ = \quad &15 \quad +_2 \quad 15 \quad +_2 \quad 7 \quad +_2 \quad 25 \quad +_2 \; 2 \; = 28 \end{aligned}$$

for the Grundy number of our example. It must be admitted that the rule is somewhat curious. In order to find the good move, if any, we need to supplement the rule by giving an inversion formula for the Welter function, but now that we have whetted the reader's appetite, we shall postpone this to the end of the chapter. In the example, the good move is unique, from 13 to 1.

NIM-ADDITION AND NEGATIVE NUMBERS

It is natural and necessary to extend the definition of Nim-sums to negative numbers, using the natural binary expansions of negative numbers, which begin with infinitely many 1s. In particular, the expansion of the largest negative number, -1, is an infinite string consisting entirely of 1's. We can perform the additions quickly by just adding another rule to those we gave for positive numbers:

$$-1 +_2 n = -1 - n.$$

The number appearing on both sides of this equality has the expansion "complementary" to that of n—since it plays an important role in the theory we have decided to give it a special name, $\bar{n}$, throughout this chapter. Thus

$$-3 +_2 7 = \bar{2} +_2 7 = \bar{5} = -6.$$

It is also natural to order the numbers in the lexicographic order of their binary expansions, namely

$$0 < 1 < 2 < \ldots < \ldots < -3 < -2 < -1$$

so that positive numbers precede negative ones, and to count 0 among the positive numbers. We shall make no particular use of this ordering, but the reader's thoughts will be clearer if he bears it in mind, and notes that in this system the "infinite singularity" is between -1 and 0.

A function $f(x)$ of the form

$$f(x) = \ldots((((x +_2 a) + b) +_2 c) + d) +_2 \cdots$$

is called an *animating function* (defined in terms of *a*ddition, *Nim*-addition, and preserving the *mating* function—see below). These functions have a particularly elegant theory. Note that they form a group under composition, for if f and g are two animating functions, so is the function $f(g(x))$, and the inverse function $f^{-1}(x)$, for which we have the formula

$$f^{-1}(x) = ((((\ldots x \ldots) - d) +_2 c) - b) +_2 a.$$

We define the *mating function* $(x \mid y)$ (not itself an animating function) to have the value $2^{n+1} - 1$ if x and y are congruent modulo 2^n but not modulo any higher power of 2, and to have the value -1 when $x = y$. Since this value depends only on the powers of 2 modulo which x and y are congruent, we have the important *invariance properties*:

$$(x \mid y) = (x + a \mid y + a) = (x +_2 a \mid y +_2 a) = (f(x) \mid f(y)),$$

for any number a, and so for any animating function f.

LEMMA. *We have the equalities*

$$n +_2 (n \mid 0) = n - 1, \qquad m +_2 (m \mid -1) = m + 1,$$

and

$$a +_2 b +_2 (a \mid b) = (a +_2 b) - 1.$$

Proof. Let $m + 1 = n$, so that $m = n - 1$. The binary expansions of these agree except that at the right-hand end m has $01 \ldots 1$, where n has $10 \ldots 0$, so that we obtain either from the other by Nim-adding the number with expansion $11 \ldots 1$, which is the number $(n \mid 0) = (m \mid -1)$. [In the case

$m = -1$, $n = 0$ we get an infinite string of 1's.] Now, using the invariance properties,

$$a +_2 b +_2 (a \mid b) = a +_2 b +_2 (a +_2 b \mid 0) = (a +_2 b) - 1.$$

THEOREM 78. *Any animating function f can be written in the form*

$$f(x) = x +_2 (x \mid a) +_2 (x \mid b) +_2 \ldots +_2 n,$$

for some numbers $a, b, \ldots, n$.

Proof. We need only verify that if $f(x)$ has such a form, then so do $f(x) +_2 m$, $f(x) - 1, f(x) + 1$. The first is obvious (replace n by $n +_2 m$), and the equations

$$f(x) - 1 = f(x) +_2 (f(x) \mid 0) = f(x) +_2 (x \mid f^{-1}(0))$$

$$f(x) + 1 = f(x) +_2 (f(x) \mid -1) = f(x) +_2 (x \mid f^{-1}(-1))$$

prove the remainder.

Now using the equation $(x \mid a) +_2 (x \mid a) = 0$, we can eliminate repetitions among the numbers $a, b, c, \ldots$ in the Theorem 78. When we have done this, the remaining numbers $a, b, c, \ldots$ are called the *poles* of f. The number n is called the *Nim* of f, and written $|f|$.

THEOREM 79. *An animating function f is determined by its poles to within a Nim-additive constant (and so completely determined by its poles and its Nim).*

Proof. Note that $(x \mid a)$ is positive except when $x = a$, when it is -1. So supposing $a, b, c, \ldots$ are distinct, the function

$$x +_2 (x \mid a) +_2 (x \mid b) +_2 \ldots +_2 n$$

has the same sign as $x +_2 n$ *unless and only unless* x is one of the poles $a, b, c, \ldots$. So the poles are characterised as the places where $f(x)$ does not have "the prevailing sign". There is a definite sense in which they are "the singularities" of f.

Consider for example the function

$$f(x) = (((x - 2) +_2 3) + 3) +_2 5.$$

How do we compute the form of Theorem 78?

It seems helpful to imagine the calculation being performed on a binary adding machine whose bell rings whenever we have an infinite carry (as in adding 1 to -1, or subtracting 1 from 0). Those numbers x for which the bell rings an odd number of times in the calculation are plainly the poles. In the course of a Nim-addition, even of a negative number, the bell never rings.

In the example, the subtraction of 2 will cause the bell to ring only for $x = 0$ and 1, so these are the poles for the function $x - 2$. The addition of 3 to $(x - 2) +_2 3$ will ring the bell only when

$$(x - 2) +_2 3 = -1, -2, \text{ or } -3$$

that is to say, when

$$x - 2 = \bar{0} +_2 3, \bar{1} +_2 3, \quad \text{or} \quad \bar{2} +_2 3 = \bar{3}, \bar{2}, \quad \text{or} \quad \bar{1},$$

which is when x itself is one of -2, -1, or 0. So we have

$$f(x) = x +_2 (x \mid 0) +_2 (x \mid 1) +_2 3 +_2 (x \mid -2) +_2 (x \mid -1) +_2 (x \mid 0) +_2 5,$$

which simplifies to

$$x +_2 (x \mid 1) +_2 (x \mid -2) +_2 (x \mid -1) +_2 6.$$

Note that the bell rings twice in the calculation corresponding to $x = 0$, so that 0 is a "double pole", and can be ignored. Note also that Nims are Nim-additive for composite functions—that is, for the function $h(x) = f(g(x))$ we have $|h| = |f| +_2 |g|$.

The function $g(x) = (((x + 2) +_2 3) - 1) +_2 5$ has the same poles and Nim as our example f, and so we have the identity $f(x) = g(x)$. There are many other identities deducible in this sort of way, which make it evident that the canonical form in terms of poles and Nim is superior to the forms defined by successive additions and Nim-additions. (Note that any function of the form seen in Theorem 78 is indeed an animating function.)

THE WELTER-NORM OF A FUNCTION f

For animating functions of zero Nim there is a second kind of Nim-additive norm, which we call the *Welt*, $[f]$, of f. It can be computed as follows. For the function f_a defined by

$$f_a(x) = x +_2 (x \mid a),$$

we have $[f_a] = a$. For other functions, we use the composition rule that if $h(x) = f(g(x))$ for functions of Nim zero, we have

$$[h] = [f] +_2 [g].$$

Let us see where this leads.

For the function $f_{a,b}$ defined by

$$f_{a,b}(x) = x +_2 (x \mid a) +_2 (x \mid b), .$$

we have the composition formula

$$\begin{aligned} f_{a,b}(x) &= x +_2 (x \mid a) +_2 (f_a(x) \mid f_a(b)) \\ &= f_a(x) +_2 (f_a(x) \mid c), \text{ say,} \\ &= f_c(f_a(x)), \end{aligned}$$

where $c = f_a(b) = b +_2 (a \mid b)$. So we have, by composition,

$$[f_{a,b}] = c +_2 a = a +_2 b +_2 (a \mid b) = (a +_2 b) - 1,$$

which we asserted before was the Welter function $[a \mid b]$.

Applying the same technique to

$$f_{a,b,c}(x) = x +_2 (x \mid a) +_2 (x \mid b) +_2 (x \mid c)$$

(for any c), we find

$$\begin{aligned} [f_{a,b,c}] &= f_{a,b}(c) +_2 [f_{a,b}] \\ &= a +_2 b +_2 c +_2 (a \mid b) +_2 (a \mid c) +_2 (b \mid c). \end{aligned}$$

So we shall do an about-face, and redefine *the Welter function*

$$[a \mid b \mid c \mid \ldots] = a +_2 b +_2 c +_2 \ldots +_2 (a \mid b) +_2 (a \mid c) +_2 \ldots +_2 (b \mid c) +_2 \ldots,$$

the Nim-sum of mating functions being taken over all unordered pairs of arguments, and for *any* animating function

$$f(x) = x +_2 (x \mid a) +_2 (x \mid b) +_2 \ldots +_2 n,$$

define the *Welt of* f as the number

$$[f] = [a \mid b \mid c \mid \ldots].$$

[There is one minor irritation. Since adding a "double pole" at k makes no difference to f, we should ideally have

$$[k \mid k \mid a \mid b \mid c \mid \ldots] = [a \mid b \mid c \mid \ldots],$$

but unfortunately we have instead

$$[k \mid k \mid a \mid b \mid c \mid \ldots] = -1 +_2 [a \mid b \mid c \mid \ldots].$$

So the value of $[f]$ depends slightly on the way that f is presented, and really we should regard the Welt of f as a pair of values $n, \bar{n}$, related by Nim-adding -1. We ignore this from now on.]

Now for distinct non-negative integers $a, b, c, \ldots$ the function $[a \mid b \mid c \mid \ldots]$ is indeed computed by the curious rule we gave at the start of the chapter. To see this, suppose again that a and b are a best-mated pair—that is to say, that they are congruent to each other modulo the highest possible power of 2. Then for any other number of the set, c, say, we have $(a \mid c) = (b \mid c)$, since a and b will be congruent to c modulo the same power of 2. So in the formula for $[a \mid b \mid c \mid \ldots]$ the terms $(a \mid c)$ and $(b \mid c)$ will cancel for all such c, and so we have the splitting

$$\begin{aligned}[a \mid b \mid c \mid \ldots] &= a +_2 b +_2 (a \mid b) +_2 c +_2 \cdots +_2 (c \mid d) +_2 \cdots \\ &= [a \mid b] +_2 [c \mid d \mid \ldots],\end{aligned}$$

which, together with the formula $[a \mid b] = (a +_2 b) - 1$, suffices to prove the rule.

So the Welter function $[a \mid b \mid c \mid \ldots]$, although it can be defined as a completely symmetric function of its arguments, nevertheless splits naturally into Welter functions of at most two arguments. It is because the properties of $(a \mid b)$ produce this mating that we call it the mating function.

Note that the Welter function is an animating function of each of its arguments. In fact, since the typical animating function

$$f(x) = x +_2 (x \mid a) +_2 (x \mid b) +_2 \cdots +_2 n$$

can also be written

$$f(x) = [x \mid a \mid b \mid c \mid \ldots] +_2 n_1,$$

where $n_1 = n +_2 [a \mid b \mid c \mid \ldots]$, the Welter function is in a sense merely the most general animating function.

There is another way of evaluating the Welter function of $k \geqslant 2$ arguments, by reducing it to functions for $k - 1$ and $k - 2$ arguments using the formula

$$[a \mid b \mid c \mid d \mid \ldots] = [[a \mid c \mid d \mid \ldots] \mid [b \mid c \mid d \mid \ldots]] +_2 [c \mid d \mid \ldots]$$

which follows immediately on expanding both sides in terms of the definition, and using the invariance property of the mating function to show that

$$([a \mid c \mid d \mid \ldots] \mid [b \mid c \mid d \mid \ldots]) = (a \mid b).$$

LEMMA. *We have* $[a' \mid b \mid c \mid \ldots] = [a \mid b' \mid c \mid \ldots]$ *if and only if*

$$[a' \mid b' \mid c \mid \ldots] = [a \mid b \mid c \mid \ldots].$$

Proof. When we expand both sides of the first equation by the formula above, we find it equivalent to

$$[a' \mid c \mid d \mid \ldots] +_2 [b \mid c \mid d \mid \ldots] = [a \mid c \mid d \mid \ldots] +_2 [b' \mid c \mid d \mid \ldots]$$

while the second equation similarly becomes

$$[a' \mid c \mid d \mid \ldots] +_2 [b' \mid c \mid d \mid \ldots] = [a \mid c \mid d \mid \ldots] +_2 [b \mid c \mid d \mid \ldots],$$

which asserts the same thing.

THE EVEN ALTERATION THEORY

We are approaching one of the most remarkable properties of the Welter function. Let us write

$$\begin{bmatrix} a & b & c & \\ a' & b' & c' & \cdots \end{bmatrix} \begin{matrix} n \\ = \\ n' \end{matrix}$$

to mean that the equation $[a \mid b \mid c \mid \ldots] = n$ remains true whenever we replace any *even* number of the letters $a, b, c, \ldots, n$ by the corresponding primed letters $a', b', c', \ldots, n'$.

THEOREM 80. *Let $[a \mid b \mid c \mid \ldots] = n$, and let n' be any number distinct from n. Then there are unique numbers $a', b', c', \ldots$ distinct respectively from the corresponding numbers $a, b, c, \ldots$, so that we have*

$$\begin{bmatrix} a & b & c & \\ a' & b' & c' & \cdots \end{bmatrix} \begin{matrix} n \\ = \\ n' \end{matrix}$$

Proof. Since the Welter function is animating in each argument, we can uniquely solve the equations

$$[a' \mid b \mid c \mid \ldots] = [a \mid b' \mid c \mid \ldots] = [a \mid b \mid c' \mid \ldots] = \ldots = n'$$

for the numbers $a', b', c', \ldots$, which will automatically be distinct from the corresponding $a, b, c, \ldots$ Since the lemma then shows, for instance, that $[a' \mid b' \mid c \mid d \mid \ldots] = n$, it provides an inductive proof of the theorem.

LEMMA. *If $b, c, d, \ldots$ are distinct, then the Welter function $[a \mid b \mid c \mid \ldots]$ has the same sign as the Nim-sum $a +_2 b +_2 c +_2 \ldots$, if and only if a is distinct from each of $b, c, d, \ldots$.*

Proof. This follows immediately from the expansion of $[a \mid b \mid c \mid \ldots]$, and the fact that $(a \mid x)$ is negative if and only if $a = x$. Recall that 0 is counted as positive.

Now to show that $[a \mid b \mid c \mid \ldots]$ is indeed the Welter function of the appropriate position in Welter's game, we must show that it is the least number from $0, 1, 2, \ldots$ which is not the value of any of the numbers

$$[a' \mid b \mid c \mid \ldots], [a \mid b' \mid c \mid \ldots], [a \mid b \mid c' \mid \ldots], \ldots \text{(the } excludents\text{)}$$

for which the arguments in each case are distinct and positive (counting 0), and $a' < a$, $b' < b$, $c' < c, \ldots$.

Now the lemma assures us that its value is positive for distinct positive $a, b, c, \ldots$, and moreover, that if n and n' are positive, and $b, c, \ldots$ distinct and positive, then the solution a' of the equation $[a' \mid b \mid c \mid \ldots] = n'$ is positive. Moreover, we know that the Welter function changes when we change any variable, so that $[a \mid b \mid c \mid \ldots]$ is certainly distinct from all its excludents. So it will suffice to prove that if $n' < n$ in the equation

$$\begin{bmatrix} a & b & c & \\ a' & b' & c' & \cdots \end{bmatrix} = \frac{n}{n'}$$

then an *odd* number of the numbers $a', b', c', \ldots$ are less than the corresponding numbers $a, b, c, \ldots$ (for this will ensure that at least one is).

LEMMA. *If*

$$\begin{bmatrix} a & b & c & \\ a' & b' & c' & \cdots \end{bmatrix} = \frac{n}{n'}$$

then we have $(a' \mid b) = (a \mid b')$, $(a \mid b) = (a' \mid b')$, *and, for any* x, *an even number of the inequalities*

$$a +_2 a' +_2 x < x$$

$$b +_2 b' +_2 x < x$$

$$c +_2 c' +_2 x < x$$

$$\cdots$$

$$n +_2 n' +_2 y < y$$

where y *is* 0 *or* x *according as the number of arguments in the Welter functions is even or odd.*

Proof. The first assertions follow easily from the invariance property of the mating function and the formula we gave for the Welter function of k arguments in terms of functions of $k - 1$ and $k - 2$ arguments.

For the remaining assertion, we suppose without loss of generality that $(a \mid b)$ is the largest mating function in the expansion of any of the given Welter functions. It then follows that the equation

$$[a \mid b] +_2 [c \mid d \mid \ldots] = n$$

remains true whenever any even number of primes are attached to $c, d, \ldots, n$, so that we have

$$\begin{bmatrix} c & d & \\ c' & d' & \cdots \end{bmatrix} = \frac{m}{m'}$$

where $m = n +_2 [a \mid b]$, $m' = n' +_2 [a \mid b]$. By the inductive hypothesis there are evenly many valid inequalities among

$$c +_2 c' +_2 x < x$$
$$d +_2 d' +_2 x < x$$
$$\dots$$
$$m +_2 m' +_2 y < y$$

the last of which is equivalent to $n +_2 n' +_2 y < y$. So the lemma is true provided evenly many of

$$a +_2 a' +_2 x < x$$
$$b +_2 b' +_2 x < x$$

are valid. But since $(a \mid b) = (a' \mid b')$ we have $[a \mid b] = [a' \mid b']$ by choice of a and b, so $a +_2 b = a' +_2 b'$, and the two given inequalities are identical.

Now we come to the last part of the even alteration theory.

THEOREM 81. *If $a, b, c, \dots, a', b', c', \dots$ are distinct and positive (counting 0), then the number of valid inequalities among*

$$a' < a, b' < b, c' < c, \dots, n' < n$$

is even.

Proof. The proof uses several identities which were used by Welter to *define* his function(!), namely

$$[0 \mid a \mid b \mid \dots] = [a-1 \mid b-1 \dots]$$
$$[a +_2 x \mid b +_2 x \mid \dots] = [a \mid b \mid c \mid \dots] +_2 y$$

where in the second case, y is 0 or x according as the number of arguments in the Welter function is even or odd. These are easily verified from our formula for his function and properties of the mating function.

Now that we do is use these identities to show that the parity of the number of valid inequalities above is the same in all cases, inductively.

Now if, say,

$$\begin{bmatrix} 0 & a & b & c & \\ x & a' & b' & c' & \end{bmatrix} \dots = \begin{matrix} n \\ n' \end{matrix}$$

then we have

$$\begin{bmatrix} a-1 & b-1 & c-1 & \\ a'-1 & b'-1 & c'-1 & \end{bmatrix} \dots = \begin{matrix} n \\ n' \end{matrix}$$

and the parities in these two cases are the same since we cannot have $x < 0$. So it suffices to show that the two equations

$$\left[\begin{array}{c|c|c|c} a & b & c & \\ a' & b' & c' & \end{array} \cdots \right] = \begin{array}{c} n \\ n' \end{array}$$

$$\left[\begin{array}{c|c|c|c} a + x & b + x & c + x & \\ a' + x & b' + x & c' + x & \end{array} \cdots \right] = \begin{array}{c} n + y \\ n' + y \end{array}$$

yield the same parity. But we know that an even number of the three inequalities

$$a' < a, a' +_2 x < a +_2 x, a +_2 a' +_2 x < x$$

are valid (the theory of Nim shows that, more generally, if $p +_2 q +_2 \cdots = s$, then an even number of the inequalities $p +_2 t < p, q +_2 t < q, \ldots, s +_2 t < s$ are valid). So it suffices to show that an even number of

$$a +_2 a' +_2 x < x$$

$$b +_2 b' +_2 x < x$$

$$\cdots$$

$$n +_2 n' +_2 y < y,$$

which is what the lemma gives.

THEOREM 82. (Welter's theorem.)' *$[a \mid b \mid c \mid \ldots]$ is the Grundy number of the corresponding position in Welter's game. Moreover, if $[a \mid b \mid c \mid \ldots] = n$, and $n' < n$, then an odd number of legal moves are available in Welter's game to take the position to one of Grundy number n', while if $n' > n$, there will be an even number (possibly zero) of such moves.*

This theorem has already been proved in the course of the previous discussion.

AN INVERSION ALGORITHM

The reader who wishes to play the game will find himself in need of an algorithm to solve equations such as

$$[a \mid b \mid c \mid \ldots \mid x] = n.$$

We first show that no such algorithm is needed if he wants only to play a single game in which there are at most four coins. For if there are exactly 4 coins, at a, b, c, d with mates a, b and c, d, then the position is a second player win if and only if

$$[a \mid b] +_2 [c \mid d] = 0,$$

that is to say, if and only if

$$[a \mid b] = [c \mid d],$$

or finally, if and only if

$$a +_2 b = c +_2 d,$$

so that the outcome of any 4-coin position is the same as that in Nim. The three coin positions have the same outcomes as Nim-positions if we number the squares from 1 instead of 0, for we can imagine a fourth coin at 0. The two coin second player wins are $2n$, $2n + 1$.

These observations can be proved without developing the general theory, and they seem to have been made again and again by many people independently.

The following seems quite a good algorithm for inverting the Welter function. If we wish to solve

$$[a \mid b \mid c \mid \ldots \mid x] = n,$$

make some hypothesis about the marital status of x (that is, whether x is the spinster, or which of $a, b, c, \ldots$ is its mate). This hypothesis enables us to complete the entire mating pattern, and enables us to solve for x. If the result confirms our hypothesis, x is the solution. If not, the new value of x is used to suggest a better hypothesis. It can be shown that the process converges after a number of steps which is bounded by both the number of binary digits in the final answer, *and* two more than half the number of arguments in the Welter function. It often converges much faster.

We take as an example the equation $[2 \mid 3 \mid 5 \mid 7 \mid x] = 0$. It seems plausible in general that a good first hypothesis is that x is ill mated—in this case that x is the spinster. This gives

$$[3 \mid 7] +_2 [2 \mid 5] +_2 x = 0,$$

whence $x = 5$. This is *very* well mated with 5, so we suppose

$$[x \mid 5] +_2 [3 \mid 7] +_2 2 = 0$$

which yields $[x \mid 5] = 5$, $x +_2 5 = 6$, *so* $x = 3$. This is now very well mated with 3, so we suppose

$$[x \mid 3] +_2 [5 \mid 7] +_2 2 = 0,$$

whence $[x \mid 3] = 3$, $x +_2 3 = 4$, $x = 7$. This finally yields

$$[x \mid 7] +_2 [3 \mid 5] +_2 2 = 0,$$

and so $[x \mid 7] = 7$, $x +_2 7 = 8$, and $x = 15$. The example has been selected to illustrate a slowly converging case, and our initial assumption was suspect—plainly x must be better mated with one of 2 and 5 than these are with each other.

If we try instead the equation

$$[2\,|\,3\,|\,5\,|\,7\,|\,11\,|\,x] = 0,$$

the initial assumption that x mates with 2 yields $x = 5$, and then the assumption that x mates with 5 yields $x = 9$, which is correct. Of course when actually playing the game we must decide which one of $a, b, c, \ldots$ to change, if we wish to decrease n to n'. I do not know of any rule which helps us to do this. However, it might be helpful to remark that the largest power of 2 dividing $n - n'$ is the same as that dividing $a - a', b - b', \ldots$, etc. This at least helps us to make sensible hypotheses about the mating behaviour.

It should be noted that the rules we have given for computing and inverting the Welter function have been chosen with mental computation in mind, so that our reader can make almost instant responses at the gaming table. The iterative technique for inversion naturally has the property that mistakes made in the initial iterations are irrelevant, and that the final answer has been checked by actually computing the Welter function.

Other algorithms for computing and inverting the Welter function are given in *Winning Ways*, where the misère form is also analysed. It turns out that Welter's game is tame!

HACKENBUSH (UNRESTRAINED)

This has also been called *Hewitt, Graph and Chopper*, and (when played with pictures of people) *Lizzie Borden's Nim*. It is played with a picture, perhaps like that of Fig. 38. The graph may have loops (the apples on the

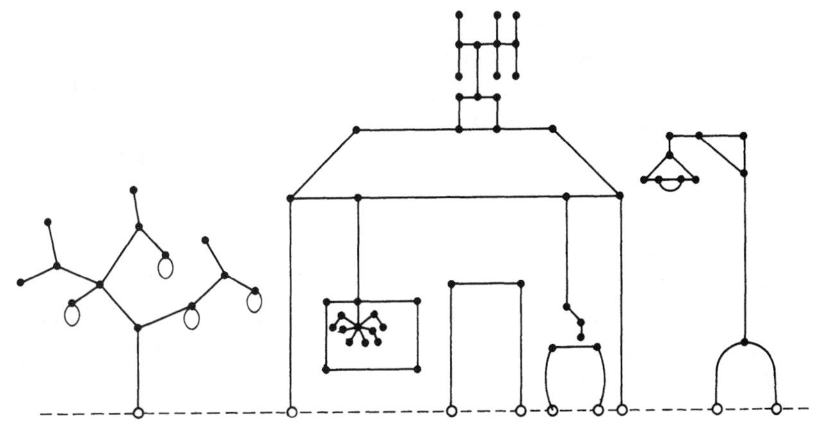

FIG. 38. The Hackenbush Estate. (Enquiries: Hackenbush, Welter and Prune.)

tree) and multiple edges (the lamp-bulb). Each component of the graph is required to contain a *base-node*—that is, to say, one of the nodes indicated in our Figure by small circles lying on a dotted line which is *not* part of the graph (and is called *the ground*).

The two players play alternately, a move consisting in the removal (chopping) of a single edge together with any part of the new graph which no longer contains a base node. So for instance, removal of the upper part of the spider's thread disposes of the spider and window—removal of the lower part disposes only of the spider. The player loses who is unable to move because no edge remains.

THE WEIGHT OF A PICTURE

We show how to compute a number, called the *weight* of a picture, which will turn out to be its Grundy number. We allow ourselves to identify any set of base nodes, or the nodes of any circuit, an edge which joined two identified nodes becoming a loop. Thus Fig. 38 has the same weight as

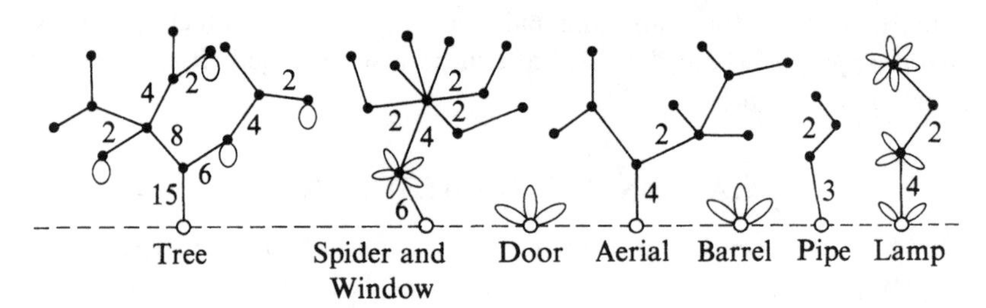

FIG. 39. The Hackenbush Garden.

Fig. 39, in which this identification has been performed, and which we have further simplified by omitting the fourteen edges of the house-frame.

Now we observe that in play, a loop at a node has just the same effect as a twig at that node, so that we shall consider the resulting diagram as if it were made out of (mathematical) *trees*. We proceed down these trees, marking any edge with the number $(a +_2 b +_2 \ldots) + 1$, where a, b, c,... are the numbers marked at the edges it immediately supports. At a twig, there is no supported edge, and so the mark is 1, and of course a similar remark holds for a loop.

These numbers we call the *stresses* at the various edges—they have been marked in Fig. 39, except that we have omitted the marks on edges of stress 1. The weight of the picture is the Nim-sum of the stresses at all edges which meet the ground (in the identified version).

With a little practice, the stresses can be inserted directly on the original

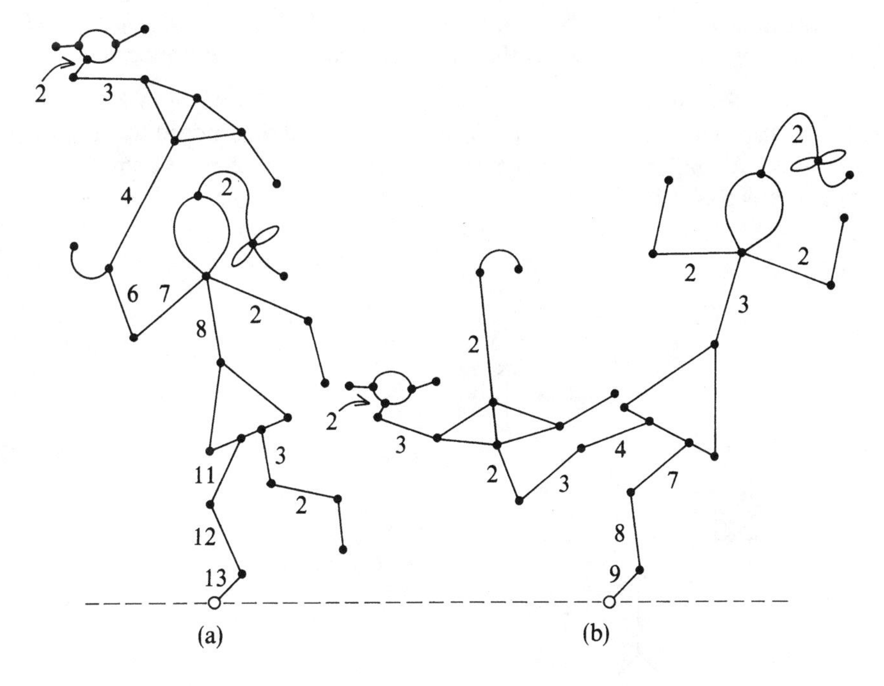

FIG. 40. Girl with umbrella and bird.

form of the diagram. In Fig. 40 we give some moderately complicated examples for the reader to test his skill. In what follows, we write $\sigma(x)$ for the stress on the edge x. This is to be carefully distinguished from another number associated with any edge—we use $\omega(x)$ for the number $(\sigma(x)\,|\,0)$, which we call the *weight* of x. As the next Theorem shows, the *weight* is the contribution of x itself, while the *stress* is that of the edge together with its load. (The *load* of an edge is the collection of edges which would disappear if that edge were deleted.)

THEOREM 83. (The weight theorem.) *The stress on any edge x is the Nim-sum of the weights of that edge and all the edges in its load.*

Proof. We consider the identified form of the picture. Then every edge other than x in the load of x is one of the edges immediately supported by x, or in the load of just one of these edges y. So by induction, the Nim-sum of the weights of all edges (other than x) in the load of x is the Nim-sum of the stresses on these edges y immediately supported by x. But this is just $\sigma(x) - 1$, by the definition of $\sigma(x)$, and Nim-adding the weight $(\sigma(x)\,|\,0)$ of x itself, we obtain the stress $\sigma(x)$.

THEOREM 84. (The change of grip theorem.) *If the loads of two edges x and y in pictures P and Q are isomorphic, then the weights of x and y are equal.*

Thus in Fig. 40(a), the stress on the girl's forearm is 6, while in Fig. 40(b) that on her foot is 2, but we see $(6 \mid 0) = (2 \mid 0) = 3$, so that indeed the weights are equal. A similar situation occurs as the strong-man changes arms in Fig. 41 and it is this that we take as our example.

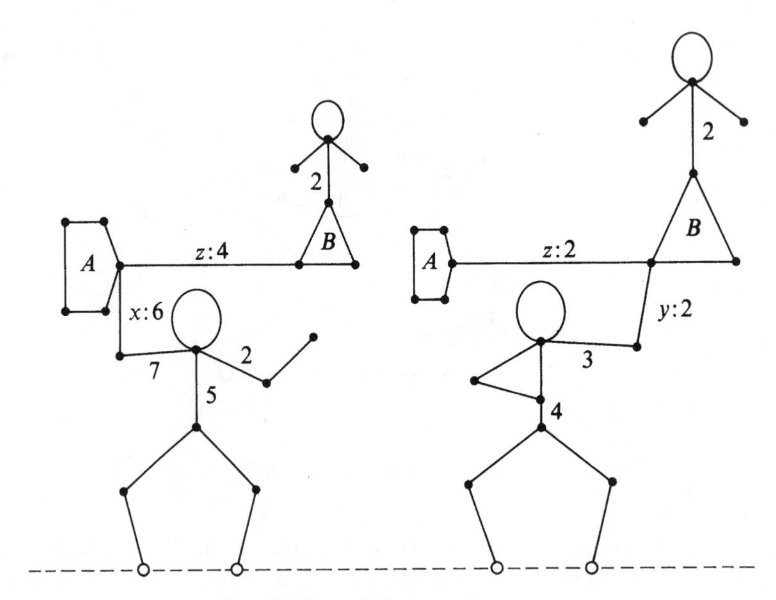

FIG. 41. The ambidextrous strong-man.

Proof. Let L be the load of x in P (or of y in Q). Since L is connected it suffices to consider the case when the endpoints of x and y lying in L are joined by an edge z. If L remains connected after deletion of z, then the load of any edge of L in P is the same as that of the corresponding edge in Q, so that $\sigma(x) = \sigma(y)$, whence $\omega(x) = \omega(y)$.

Otherwise deletion of z from L results in two components A and B, whose weights a and b are the same in P as they are in Q. (In Fig. 41, $a = 1$, $b = 3$.) Then the stresses on x and y are

$$\sigma(x) = ((b + 1) +_2 a) + 1, \qquad \sigma(y) = ((a + 1) +_2 b) + 1,$$

and using the invariance property of the mating function,

$$\begin{aligned}\omega(x) &= (((b + 1) +_2 a) + 1 \mid 0) \\ &= ((b + 1) +_2 a \mid -1)\end{aligned}$$

$$= (b + 1 \mid -a - 1)$$
$$= (a + 1 \mid -b - 1)$$
$$= \omega(y)$$

similarly, proving the theorem.

Now suppose some picture P is supplemented by adjoining a new piece of weight ω at some point of P, so as to obtain a new picture, which we shall call $P(\omega)$. How does the weight of $P(\omega)$ vary with ω? We assert that it is plain from the definition of the weight of a picture that the result is in fact an animating function of ω. The same holds for the stress on any edge in P as ω varies. It follows that when we chop an edge x, the stresses on edges y supporting x change in accordance with the following law: if y supports x, and $\sigma_x(y)$ denotes the stress on y after x is deleted, then we have

$$(\sigma(y) \mid \sigma_x(y)) = \omega(x).$$

Now we define a *cycle* as a set of distinct edges forming a circuit, or a path connecting the base to itself. We say that two edges are *concyclic* if they belong to exactly the same non-empty set of cycles.

In Fig. 42 the legs of the boy form one class of concyclic edges, the legs of

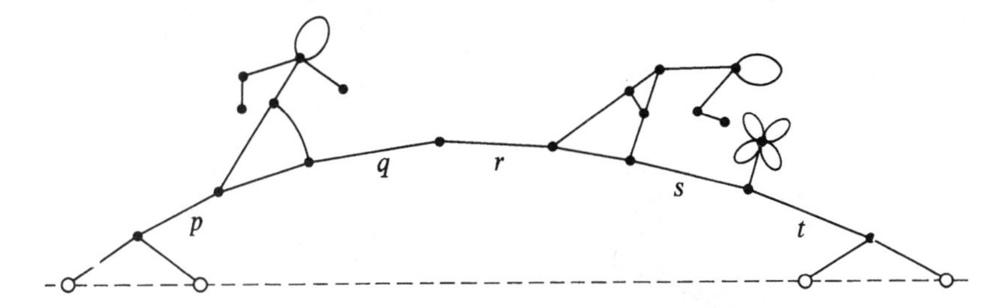

FIG. 42. The lovers' bridge.

the girl another, and the two sides of her skirt a third. The only other class containing more than one edge is $\{p, q, r, s, t\}$.

We define the function $(x \mid y) = (\sigma_x(y) \mid 0)$, the weight of y in the picture obtained by deleting x.

THEOREM 85. (The concyclic edges theorem.) *On any concyclic class of edges, the function $(x \mid y)$ has the properties (like the mating function for numbers):*

(i) $(x \mid y) = (y \mid x)$

(ii) *If x, y, z are distinct edges, then some two of the three numbers $(x \mid y)$, $(x \mid z)$, $(y \mid z)$ are equal and strictly less than the third.*

Proof. The symmetry of this function follows from Theorem 84. Let r, s, t be three concyclic edges, and consider any cycle containing them. At least one of the endpoints of the three edges is connected to the ground by a path not containing any of the three edges. If r, s, t and their endpoints r_1, r_2, s_1, s_2, t_1, t_2 are labelled in order round the cycle so that r_1 is such an endpoint, then t_2 will be another. Comparing Fig. 43(*s*) and 43(*t*) we see that

$$(\sigma_t(r) \mid \sigma_s(r)) = (t \mid s),$$

and since

$$(\sigma_t(r) \mid 0) = (t \mid r),$$

$$(\sigma_s(r) \mid 0) = (s \mid r),$$

the property (ii) for edges follows from the corresponding property for numbers.

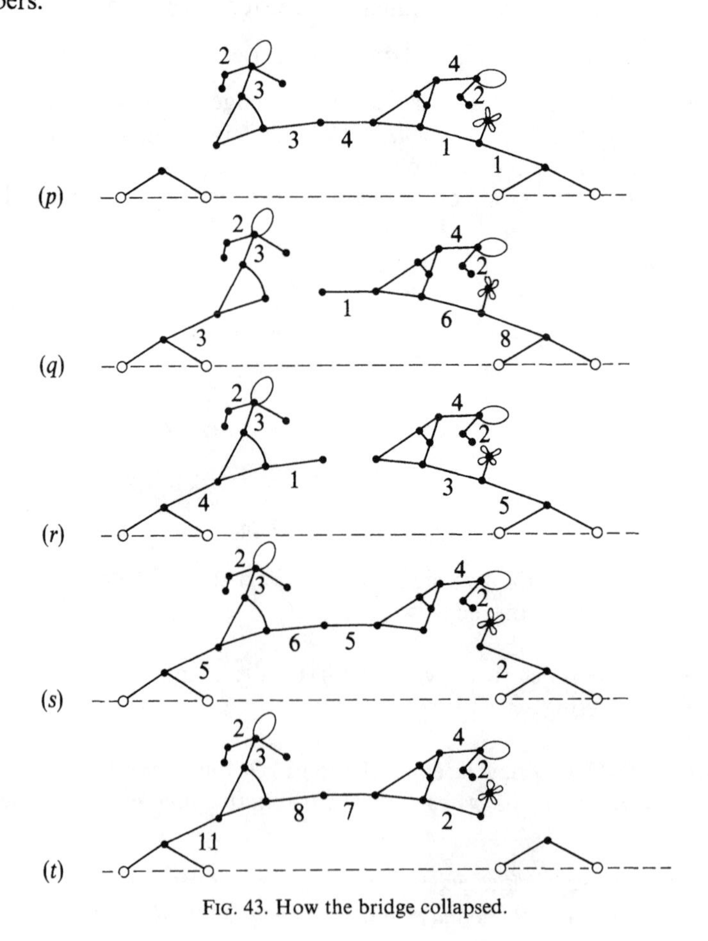

FIG. 43. How the bridge collapsed.

THEOREM 86. (The Hackenbush theorem.) *The weight $\omega(P)$ of a picture is the Grundy number of P regarded as a position in the Hackenbush game.*

Before proving this, we can use it to find the winning move in the Hackenbush Homestead (Fig. 38). Let us examine the equivalent Hackenbush Garden shown in Fig. 39. Here the total weight is

$$15 +_2 1 +_2 6 +_2 4 +_2 1 +_2 3 +_2 4 = 15 +_2 5.$$

We can therefore change this to 0 by chopping some branch of the tree so as to change the tree-trunk's stress from 15 to 5. The trunk supports two branches of stresses 8 and 6, and we can alter these to make their Nim-sum 4 by changing 8 to 2. The stress 8 branch presently supports branches of stresses 2, 1, and 4 which can be made to have Nim-sum 1 by changing 4 to 2. Climbing the tree in this way, we eventually discover the winning move (which can just as easily be proved to be unique): chop the twig bearing the highest apple on the tree!

Proof. We decompose the picture into portions P_i $(0 \leqslant i \leqslant k)$ by considering the edges that meet the ground in the identified version. Some of these edges, $x_1, \ldots, x_k$, say, support other edges, and the edge x_i and its load constitute the portion P_i $(1 \leqslant i \leqslant k)$. The portion P_0 consists of all the other edges. Defining the weight of a portion as the Nim-sum of the weights of its edges, we see that the weight of P is the Nim-sum of the weights of the P_i.

Let P' be the picture obtained from P by chopping a typical edge, and if Q is any part of P, let Q' be the part of P' consisting of edges which lay in Q, edges which have disappeared being ignored.

Now chopping an edge in P_i does not affect loads in P_j $(j \neq i)$, and so the weight $\omega(P_j)$ is unchanged, while the weight of P_i is replaced by a number $\omega(P'_1)$. So for this one move, the picture P behaves like the disjunctive sum of the portions P_i, and we need only show that $\omega(P_i)$ is the mex of all the numbers $\omega(P'_i)$ obtained by chopping edges in P_i.

Now for $i \neq 0$ we consider the picture $\overline{P_i}$ formed by the load of x_i with the upper endpoint of x_i taken as its only base-node. See Fig. 44. We can

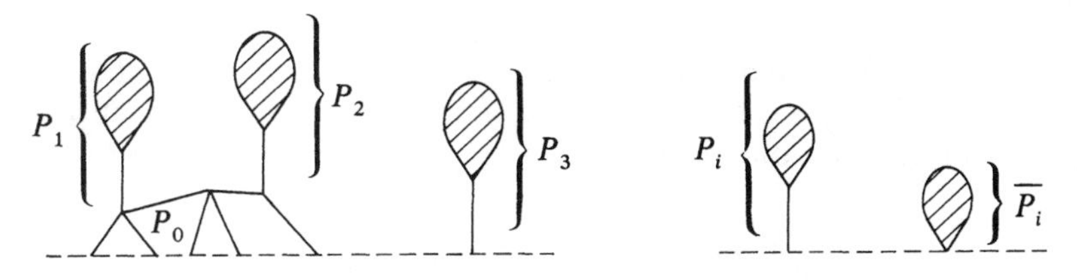

FIG. 44. The portions P_i and the definition of $\overline{P_i}$.

suppose inductively that $\omega(P_i) = \text{mex}\,\omega(P_i')$, and since each P_i' is *either* obtained from some $\overline{P}_i'$ *or* obtained by chopping x_i, we have

$$\omega(P_i) = \text{mex}\,(0, 1 + \omega(\overline{P}_i')) = 1 + \omega(\overline{P}_i),$$

which proves the desired assertion for $i \neq 0$.

For $i = 0$, we let P_0' be obtained by deleting a typical edge s from P_0, and $C = \{p, q, r, s, \ldots\}$ be the class of edges concyclic with s.

Then we have, by the definition of weight,

$$\omega(P_0') = (s \mid p) +_2 (s \mid q) +_2 (s \mid r) +_2 \ldots +_2 \Sigma_2\, 1,$$

the final Nim-sum taken over all edges of P_0 which are not in C. Since $(s \mid p) +_2 (s \mid q) +_2 \ldots$ has parity opposite to the number of edges in C, we see first that $\omega(P_0') \neq \omega(P_0)$.

Now $\omega(P_0)$ is 0 or 1 according as there are an even number or odd number of edges in P_0, so we now need only show that when $\omega(P_0) = 1$ there is an edge s so that $\omega(P_0') = 0$. But then P_0 has an odd number of edges, and so there is an odd number of edges in some concyclic class C, and for this C the final Nim-sum in the above formula for $\omega(P_0')$ is 0. Now the function $(x \mid y)$ on the edges of C induces a mating on C, taking as the first mated pair a pair x, y with maximal $(x \mid y)$, then deleting these edges and selecting the next best mated pair, and so on. If s is the spinster in this mating, then we have $(s \mid x) = (s \mid y)$ for each such pair x, y, and so the Nim-sum $(s \mid p) +_2 (s \mid q) +_2 \ldots$ is also zero, proving the result.

From *Figs* 43(p) to 43(s) we can read off the matrix of values of $\sigma_x(y)$ in the set $\{p, q, r, s, t\}$, and thence the corresponding matrix of values of $(x \mid y)$:

$\sigma_x(y)$		p	q	r	s	t
	p	–	3	4	1	1
	q	3	–	1	6	8
x	r	4	1	–	5	5
	s	5	6	5	–	2
	t	11	8	7	2	–

$(x \mid y)$		p	q	r	s	t
	p	–	1	7	1	1
	q	1	–	1	3	15
x	r	7	1	–	1	1
	s	1	3	1	–	3
	t	1	15	1	3	–

The first pair is therefore $\{q, t\}$, and then the pair $\{p, r\}$ completes $\{q, t\}$ to a quartet, leaving s as the spinster. Deleting her, we have Fig. 43(s) whose weight is indeed the combined weight of boy, node, girl, and flower.

CHAPTER 14

How to Play Several Games at Once in a Dozen Different Ways

"Home, James, and don't spare the horses!"

Since we are still concerned with impartial games, in this chapter we shall call our players Arthur and Bertha rather than Left and Right. Now there are many different ways in which Arthur and Bertha can show off by playing several games against each other simultaneously, so as to make a single compound game. Throughout much of this book we have been concerned with the *disjunctive* compound, when the compound move is defined to be a move in just one of the component games. In this chapter, we shall add some other systems of rules, so as to make in all a round dozen of different ways of playing several games at once.

Rules for moving in the compound game

(1) The *selective* (SOME) rule:
select *some* of the component games, and then make a legal move in each game you have selected.

(2) The *conjunctive* (ALL) rule:
make a legal move in *all* the component games that have not yet ended.

(3) The *disjunctive* (ONE) rule:
make a legal move in just *one* of the component games.

Rules for ending the compound game

(a) The *long* rule: the component game ends only when *all* of the component games have ended.

(b) The *short* rule: The compound game ends as soon as *any one* of the component games has ended.

Rules for deciding who wins the compound game

(+) The *normal* rule: the last player able to move is the *winner.*

(−) The *misère* rule: the last player able to move is the *loser.*

There is a very definite sense in which the normal rule is more natural than the misère rule, since it seems more sensible to agree that a player unable to move *loses* than that he *wins.*

Since we have 3 rules for moving, 2 for ending, and 2 for deciding who wins, we have indeed $3 \times 2 \times 2 = 12$ ways of playing several games at once, as we promised. For selective and disjunctive compounds, we should be able to move as long as any component remains unfinished, and so we should prefer the long ending rule, but conjunctive compounds should naturally end when any component does, and so for them the short ending rule is more natural. So we name the possibilities as follows:

$G \vee H \vee K \ldots$, the *selective* compound (*long* ending rule)

$G \triangledown H \triangledown K \ldots$, the *shortened selective* compound (*short* rule)

$G \wedge H \wedge K \ldots$, the *conjunctive* compound (*short* rule)

$G \triangle H \triangle K \ldots$, the *continued conjunctive* compound (*long* rule)

$G + H + K \ldots$, the *disjunctive* compound (*long* rule)

$G \oplus H \oplus K \ldots$, the *diminished disjunctive* compound (*short* rule)

each with either normal or misère play.

HOW CAN WE FIND OUT WHO HAS THE WINNING STRATEGY?

In any compound game of any of these types, we know that just one of the two players has a winning strategy, so that the outcome of the compound is determined in some way from the structure of the component games. Now just how much do we need to know about these component games in order to be able to compute the outcome of the compound? We know that for normal disjunctive compounds the answer to this question is "precisely the Grundy number", and that for misère disjunctive compounds the answer is much more complicated. Since neither of these answers is exactly what we should expect, the question cannot be entirely trivial. In general, we might expect some kind of "number" for each component, together with a rule for "adding" these numbers.

Recall the definition of the normal and misère *outcomes* $o^+(G)$ and $o^-(G)$— these are the symbols N or P according as it is the Next or Previous player who has the winning strategy from G (in respectively normal and misère play).

Professor C. A. B. Smith has analysed most of these compounds in a very entertaining paper [J. Comb. Theory 1 (1966) 51–81], and for the most part we have followed his analysis and names for the compounds. But some of the compounds are not analysed by Smith, and in particular, the solution of the continued conjunctive compound seems to be new. We omit the easy proofs.

WHO WINS THE SELECTIVE COMPOUND?

After playing a few selective compounds, Arthur and Bertha soon found the rule for normal play—you simply make good moves in all the games you can. In other words:

The normal outcome of $G \vee H \vee K \ldots$ *is* P *if and only if the normal outcomes of* $G, H, K, \ldots$ *are all* P.

It took them a little longer to work out the rule for misère play:

Unless all but one of the components have ended, the misère outcome of $G \vee H \vee K \ldots$ *is the same as its normal outcome. Otherwise its misère outcome is the same as the misère outcome of the only remaining component.*

WHO WINS THE SHORTENED SELECTIVE COMPOUND?

Here, if any component has ended, so has the whole game, by definition, and its outcome is P in normal play, N in misère play. Otherwise:

The normal outcome of $G \bigtriangledown H \bigtriangledown K \ldots$ *is* P *if and only if the normal outcomes of* $G, H, K, \ldots$ *are all* P,

and similarly

The misère outcome of $G \bigtriangledown H \bigtriangledown K \ldots$ *is* P *if and only if the misère outcomes of* $G, H, K, \ldots$ *are all* P.

So again we need only know who wins the individual components, and which components have ended. In fact the rule is even simpler than for the ordinary selective compound, since the rule for misère play does not now need the outcomes under normal play.

WHO WINS THE CONJUNCTIVE COMPOUND?

When Arthur knows that he can win a game he is playing with Bertha, he usually tries to beat her as quickly as possible, so that he can boast of having won the game in very few moves. Bertha conversely tries to postpone her defeat as long as possible. Now when a game is played in this way between intelligent players (the winner trying to win quickly and the loser to lose slowly), it always lasts for exactly the same number of moves, and this number is called the *remoteness* of the game. Professor Steinhaus, who invented this concept, gave rules for calculating the remotenesses of the various positions of a game under normal play, as follows:

(i) If G has an option of even remoteness, the remoteness of G is *one more* than the *minimal even* remoteness of any option of G.

(ii) If not, but G has options of odd remoteness, then the remoteness of G is *one more* than the *maximal odd* remoteness of any option of G.

(iii) The remoteness of an ended position is *zero*.

For normal play the P-positions are those of even remoteness, so rule (i) corresponds to the winner's aim of shortening the game, rule (ii) to the loser's of lengthening it, and rule (iii) to the fact that a game with no possible move lasts for zero time. We write $R^+(G)$ for the remoteness of G under normal play.

There are similar rules for remotenesses under misère play, but since then P-positions have odd remoteness, we must interchange the words *odd* and *even* in the above rules. The remoteness under misère play we call $R^-(G)$.

REMOTENESS OF CONJUNCTIVE COMPOUNDS

It soon dawned on Arthur and Bertha that when they played conjunctive compounds the game only lasted as long as the shortest component, for the winner of that component could always exercise delaying tactics in the others. In other words:

The remoteness of a conjunctive compound is the same as the minimal remoteness of any of the component games.

This applies to both the normal and misère remotenesses. So to work out who wins a conjunctive compound, we need only know the appropriate remotenesses of the individual components, and we can then see whether the smallest one of these is even or odd.

WHO WINS THE CONTINUED CONJUNCTIVE COMPOUND?

Bertha's winning tactics are subtly different from Arthur's. When she knows that she can win, she enjoys prolonging Arthur's agony, and tries to make the game last as long as possible! Arthur, conversely, prefers to end the game quickly, since he hates to play a game he cannot win. Now when a game is played in this way between intelligent players (the winner trying to win slowly and the loser to lose quickly), the number of moves it lasts is called the *suspense* number of the game. Rules for calculating this number for normal play are as follows:

(i) If G has an option of even suspense number, the suspense of G is *one more* than the *maximal even* suspense number of any option of G.

(ii) If not, but G has an option of odd suspense number, the suspense of G is *one more* than the *minimal odd* suspense number of any option of G.

(iii) The suspense number of an ended position is *zero.*

In other words, we interchange the words *minimal* and *maximal* in Professor Steinhaus's rules. For calculating suspense numbers in misère play we would also interchange *even* and *odd* again. We shall use $S^+(G)$ and $S^-(G)$ for the suspense numbers of G under respectively normal and misère play.

SUSPENSE NUMBERS OF CONTINUED CONJUNCTIVE COMPOUNDS

Arthur dislikes playing continued conjunctive compounds, since they last so long, and Bertha usually wins. The reason is that she worked out the rule by analogy with that for the ordinary conjunctive compounds:

The suspense number of a continued conjunctive compound is the same as the maximal suspense number of any of the component games.

Once again this applies in both normal and misère play. The idea is that the winner of the game with largest suspense number can certainly delay the end of the compound until this particular game ends, and during this time she can have disposed of any of the components she is forced to lose. So to find out who wins a continued conjunctive compound, we need only know the suspense numbers of the component games, and decide whether the largest of these is even or odd.

WHO WINS THE DISJUNCTIVE COMPOUND?

We already know how to work out the outcome of a disjunctive compound under normal play. For completeness we repeat it here:

The Grundy number of $G + H + K\ldots$ is the Nim-sum of the Grundy numbers of the component games $G, H, K, \ldots$. The normal outcome is P if and only if the Grundy number is 0.

We also repeat the rule for computing Grundy numbers.

(i) The Grundy number of G is the *least* number (from 0, 1, 2, ...) which is not the Grundy number of any option of G.

In particular

(ii) The Grundy number of any ended position is *zero*.

We call the Grundy number so defined the *normal Grundy number*, $G^+(G)$, and we have also defined the *misère Grundy number* $G^-(G)$, by replacing *zero* by *one* in rule (ii), and so making rule (i) apply only to games that have not yet ended. For complicated games the misère and normal Grundy numbers can be quite unrelated. But the particular game of Nim has the property that in every position *either* the two Grundy numbers are equal *or* one of them is 0 and the other is 1. If this also holds for all positions of some game G, we call G *tame*.

The disjunctive compound $G + H + K \ldots$ of any number of tame games is tame, and then the two Grundy numbers of $G + H + K \ldots$ are equal if and only if the two Grundy numbers of some one of G, H, K, ... are equal.

So to work out who wins a disjunctive compound of tame games we need only know the two Grundy numbers (normal and misère) for each component. From the normal Grundy numbers of the components we Nim-add to find the normal Grundy number of the compound. The misère Grundy number is either the same as this, or the two numbers are 0 and 1, and the latter possibility will only happen for the compound when it happens for every component. Since the misère outcome of a game is P if and only if the misère Grundy number is 0, the rule suffices to find misère outcomes of compounds of tame games.

Another way of remembering the rule is to note that for any tame game there is a Nim-position with the same pair of Grundy numbers. We can then replace each component by the appropriate Nim-position (which might well have more than one heap) and pretend we are playing misère Nim. But for non-tame games we must read Chapter 12.

WHO WINS THE DIMINISHED DISJUNCTIVE COMPOUND?

Both the rules for diminished disjunctive compounds are easier than the misère rule for ordinary disjunctives. The idea is that we must pay special attention to positions near the end of the game. What we do is *foreclose* the game by making a position illegal if the game has just ended, or can be ended by a single *winning* move. Modern Chess is in fact the foreclosed version of primitive Chess, in which the game ended when the loser's King was captured. In modern Chess, a position in which the King has been captured is illegal, as is also any position in which the King can be captured on the next move, and so a game ends when one of the Kings is checkmated.

Of course the result of foreclosing a game in misère play will probably not be the same as that of foreclosing it under normal play, since the notions of winning moves differ. We now define the (normal and misère) *foreclosed Grundy numbers*, $F^+(G)$ and $F^-(G)$, to be the *normal* Grundy numbers of the two foreclosed versions of G. Of course these numbers will not be defined if G has already ended or can be ended in one winning legal move, for then the foreclosed game has no legal positions and will not exist. But for such games we shall still need to know the outcome. Then the rule for diminished disjunctive compounds is:

The foreclosed Grundy number of $G \oplus H \oplus K \ldots$ is defined if and only if those of $G, H, K, \ldots$ are all defined, and is then their Nim-sum. The outcome is P if and only if the foreclosed Grundy number is 0, or some component has outcome P but undefined foreclosed Grundy number.

In other words, if one of the components has ended, or can be ended in a single winning move, the same is true of the compound. Otherwise the foreclosed compound is the same as the normal play disjunctive compound of foreclosed components. Of course, the foreclosed Grundy numbers we use are the normal ones for normal play, and the misère ones for misère play.

In the

selective	shortened selective	conjunctive	continued conjunctive	disjunctive	diminished disjunctive

compound,

$G \vee H \ldots$	$G \nabla H \ldots$	$G \wedge H \ldots$	$G \triangle H \ldots$	$G + H \ldots$	$G \oplus H \ldots$

of a number of games $G, H, \ldots$, we *move* in

some	some	all	all	one	one

of those components which have not yet ended, and the game *ends* as soon as

each	any	any	each	each	any

of the component games has ended. The solution for normal play uses

o^+	o^+, end?	R^+	S^+	G^+	F^+, o^+

while that for misère play uses

o^+, o^-, end?	o^-	R^-	S^-	G^+, G^-, tame?	F^-, o^-.

FIG. 45. Tactics for a dozen different ways of playing several games at once.

It will be seen from Fig. 45 that in eleven of the twelve cases the solution really involves nothing worse than computing a simple numerical function for each component, from which the outcome of the compound can be easily calculated. The same is true for misère play disjunctive compounds only with the condition of tameness. If his games are not tame, the innocent reader would be wise to refuse to play them, but a more foolhardy reader will be eager to use the more difficult theory of Chapter 12.

VARIATIONS ON THE THEME

In this chapter we have supposed that the games are impartial in the sense that any move which is legal for one player is also legal for the other. Most of this book is the result of the attempt to remove this restriction for the case of normal play disjunctive compounds. It is removed for the normal play selective compounds in Chapter 10 of *Winning Ways*, which also discusses some other variations. Of course we can always regard any game as impartial by "building in" any restrictions on possible moves into the position, so that in Chess, for example, each position would be regarded as carrying with it the colour of the next piece to be moved. But this has the unfortunate effect that if Chess were a component of some compound game a player might find himself moving differently coloured pieces at different times. For the conjunctive compounds, such problems do not arise, since the move in each component automatically alternates.

We have also supposed that each game has only a finite number of positions (i.e., is a *short* game). It is perfectly possible to replace this by the condition used elsewhere in the book that the game lasts for a finite, but possibly unbounded number of moves, and the theories are not much altered. The curious reader will find details for some of the cases in *Winning Ways* or Professor Smith's original paper. What usually happens is that the numerical functions involved are allowed to take new values ∞ or infinite ordinal numbers, and the finite theory generalises easily.

If instead we allow a game to proceed indefinitely, an infinitely long play being counted as a draw, then the theories become rather dull except for the disjunctive compounds, which we have already considered in Chapter 11. Other conventions which permit draws can usually be converted into this one by adding new legal moves from drawn positions to themselves.

We can modify the rules about who wins and who loses, by marking the ended positions of individual components with the corresponding letters N and P. The interesting cases are the disjunctive and diminished disjunctive compounds, since in other cases several games may end simultaneously, and there is no obvious rule for deciding who wins the compound. If we define the winner of a diminished disjunctive compound to be the winner of the first component to end, then our rules will still apply, if we use the appropriate kind of foreclosed Grundy number. On the other hand, it seems clear that the theory for ordinary disjunctive compounds so generalised is much harder than the misère play theory of Chapter 12.

Finally, we can consider new systems of rules for deciding what counts as a move in the compound game. For instance, we might demand a move in just *two* components, or alternatively a move in any number strictly less than *five*. I have not been able to give a complete theory for any new rule of this type, although there is still room for hope in the case when we require a move in any *odd* number of components. Even in the absence of a general theory, one can attack the case when each component is a Nim-heap, and often we find some curious results. We discuss only one.

Moore's game "Nim_k"

Here we have a number of heaps of counters, and the move is to remove some counters from any number up to k of heaps. Ordinary Nim is the particular case Nim_1. There is a remarkable strategy in the general case:

> *Mentally split each heap into heaps whose sizes are distinct powers of 2. Then the position is P if and only if the numbers of heaps of each size are all divisible by $k + 1$, after this alteration.*

In other words, we write the numbers in the binary notation, but then add these numbers without carry, and in the scale of $k + 1$, and the position is P if and only if the resulting "number" is zero!

PLAYING SEVERAL DIFFERENT GAMES IN SEVERAL DIFFERENT WAYS AT ONCE

It is possible to play a selective compound of games which are themselves conjunctive compounds (say) of smaller games. Is there any way of telling how to win such compounded compounds? The only easy cases are those with selective compounds outermost (since their outcomes depend only on the outcomes of individual components), and certain combinations of selective and shortened selectives with conjunctives and continued conjunctives.

The idea is that in normal play we can compute the remoteness or suspense number of a selective or shortened selective compound from those of its components, according to the tables in Figs 46 and 47.

For misère play, there is no similar theory for ordinary selective compounds, but for the shortened selectives we have Fig. 48.

To justify these tables, note that a sensible loser of such a compound will try to move in just one component if he wants to drag things out, and in all components he can if he wants to end things quickly. So the remoteness of

$R^+(G \vee H)$

	0	1	2	3	4	5	6	7
0	0	1	2	3	4	5	6	7
1	1	1	3	3	5	5	7	7
2	2	3	4	5	6	7		
3	3	3	5	5	7	7		
4	4	5	6	7				
5	5	5	7	7				
6	6	7						
7	7	7						

$S^+(G \vee H)$

	0	1	2	3	4	5	6	7
0	0	1	2	3	4	5	6	7
1	1	1	3	3	5	5	7	7
2	2	3	2	3	4	5	6	7
3	3	3	3	3	5	5	7	7
4	4	5	4	5	4	5	6	7
5	5	5	5	5	5	5	7	7
6	6	7	6	7	6	7	6	7
7	7	7	7	7	7	7	7	7

FIG. 46.

$R^+(G \nabla H)$

	0	1	2	3	4	5	6	7
0	0	0	0	0	0	0	0	0
1	0	1	1	1	1	1	1	1
2	0	1	2	3	4	5	6	7
3	0	1	3	3	5	5	7	7
4	0	1	4	5	6	7		
5	0	1	5	5	7	7		
6	0	1	6	7				
7	0	1	7	7				

$S^+(G \nabla H)$

	0	1	2	3	4	5	6	7
0	0	0	0	0	0	0	0	0
1	0	1	1	1	1	1	1	1
2	0	1	2	3	2	3	2	3
3	0	1	3	3	3	3	3	3
4	0	1	2	3	4	5	4	5
5	0	1	3	3	5	5	5	5
6	0	1	2	3	4	5	6	7
7	0	1	3	3	5	5	7	7

FIG. 47.

$R^-(G \nabla H)$

	0	1	2	3	4	5	6
0	0	0	0	0	0	0	0
1	0	1	2	3	4	5	6
2	0	2	2	4	4	6	6
3	0	3	4	5	6		
4	0	4	4	6	6		
5	0	5	6				
6	0	6	6				

$S^-(G \nabla H)$

	0	1	2	3	4	5	6
0	0	0	0	0	0	0	0
1	0	1	2	1	2	1	2
2	0	2	2	2	2	2	2
3	0	1	2	3	4	3	4
4	0	2	2	4	4	4	4
5	0	1	2	3	4	5	6
6	0	2	2	4	4	6	6

FIG. 48.

the compound will be roughly the sum of the individual remotenesses, and the suspense function roughly the maximum for an ordinary selective compound and the minimum for a shortened selective compound. The slight divergences from this are due to the special conditions that prevail very near the beginning or end of the game. Since the misère outcome for ordinary selective compounds depends on both the normal and misère outcomes of the components, there can be no exactly similar rule for that case.

ALL THE KING'S HORSES, ETC.

There are twelve forms of this game. The game is played on a large board ruled into squares, the two by two square at the top left hand corner being called *home*. The pieces are called *horses* and move like the knights in Chess, except that only the four moves in the directions shown in Fig. 49 are allowed, and as many knights as we wish may occupy the same square.

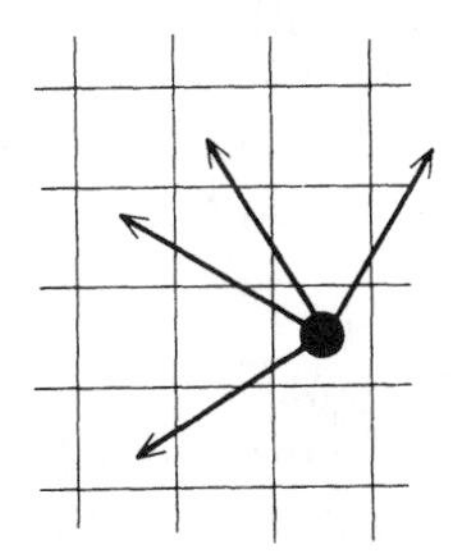

FIG. 49. The way a horse moves.

In the version:

All the King's Horses, last home loses

The player whose turn it is to move must move every horse that is not yet home, and the last player able to move in this way is the loser. It is therefore a continued conjunctive compound with misère play, the component games corresponding to the individual horses. The twelve versions range in this way from

Some of the King's Horses, first home wins

(Normal play shortened selective), to

One of the King's Horses, last home loses

(Misère play disjunctive).

We can give the winning strategies by giving tables showing the appropriate

```
P P ——P P ——P P ——          ———P P P —P P P —P P P
P P ——P P ——P P ——          ————P ————P ————P
————————————                P ————————————
————————————                P P —P —P —P —P
P P ——P P ——P P             P ————P —P —P
P P ——P P ——P P             ————P —P —P
——————                      P ————P —P
——————                      P P —P —P
P P ——P P                   P ————P
P P ——P P   o+              ————P   o−
————                        P ——
————                        P P
P P                         P
P P
        (a)                          (b)
```

FIG. 50.

functions of each square on the board, as we do on the next few pages. It turns out that the last version mentioned is not a tame game, so we do the best we can and tabulate merely the $G*$-values of the various positions, as in Chapter 12. To make our descriptions of the strategies easier, we shall suppose that every horse is removed from the board as soon as it reaches home.

Some of the King's Horses (Fig 50)

If first or last home wins, your move should leave all remaining horses in P positions in the o^+ diagram (Fig. 50(a)). If first home loses, the move should leave all horses in P positions in the o^- diagram (Fig 50(b)). If the last home loses, all horses should be in P positions in the o^+ diagram until there is only

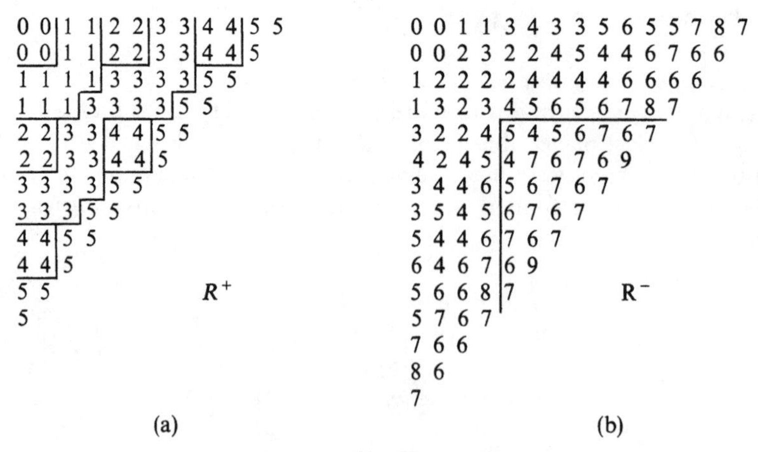

FIG. 51.

one left, which should be put into a P position in the o^- diagram. We have written — for N in these diagrams so as to make the P positions obvious.

All of the King's horses (Figs. 51, 52)

If first home wins, move so as to make the least number on the R^+ diagram (Fig. 51(a)) *even*—if first home loses, move so as to make the least number on the R^- diagram (Fig. 51(b)) *odd*. We have drawn lines partitioning the entries in these diagrams so as to make the patterns clearer. The pattern in the R^- diagram is easier to follow if we read the entries in each row in blocks of four. Similar comments apply to the S^+ and S^- diagrams.

If last home wins, move so as to make the greatest number on the S^+ diagram (Fig. 52(a)) *even*; if last home loses, the greatest number on the S^- diagram (Fig. 52(b)) should be *odd*.

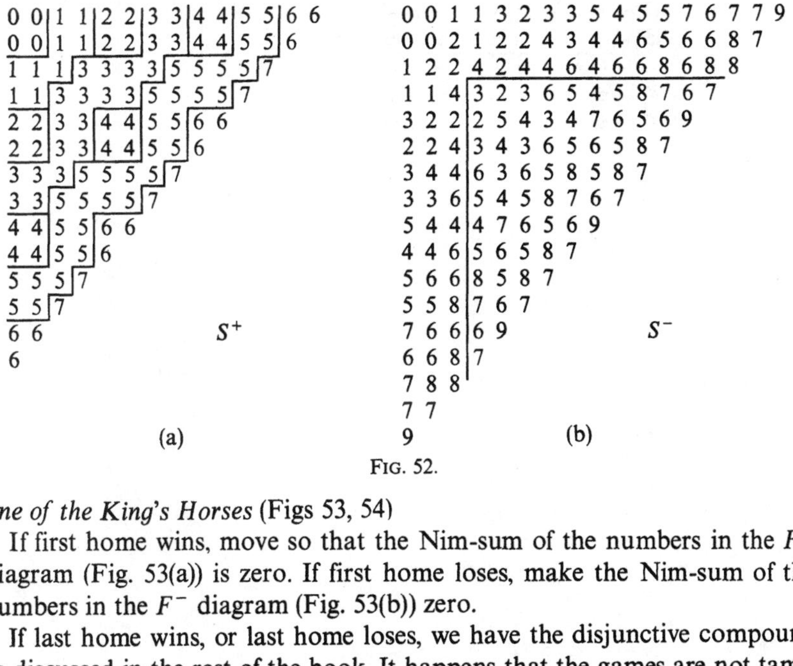

```
0 0 1 1 2 2 3 3 4 4 5 5 6 6
0 0 1 1 2 2 3 3 4 4 5 5 6
1 1 1 3 3 3 3 5 5 5 5 7
1 1 3 3 3 3 5 5 5 5 7
2 2 3 3 4 4 5 5 6 6
2 2 3 3 4 4 5 5 6
3 3 3 5 5 5 5 7
3 3 5 5 5 5 7
4 4 5 5 6 6
4 4 5 5 6
5 5 5 7
5 5 7
6 6
6
```

S^+

(a)

```
0 0 1 1 3 2 3 3 5 4 5 5 7 6 7 7 9
0 0 2 1 2 2 4 3 4 4 6 5 6 6 8 7
1 2 2 4 2 4 4 6 4 6 6 8 6 8 8
1 1 4 3 2 3 6 5 4 5 8 7 6 7
3 2 2 2 5 4 3 4 7 6 5 6 9
2 2 4 3 4 3 6 5 6 5 8 7
3 4 4 6 3 6 5 8 5 8 7
3 3 6 5 4 5 8 7 6 7
5 4 4 4 7 6 5 6 9
4 4 6 5 6 5 8 7
5 6 6 8 5 8 7
5 5 8 7 6 7
7 6 6 6 9
6 6 8 7
7 8 8
7 7
9
```

S^-

(b)

FIG. 52.

One of the King's Horses (Figs 53, 54)

If first home wins, move so that the Nim-sum of the numbers in the F^+ diagram (Fig. 53(a)) is zero. If first home loses, make the Nim-sum of the numbers in the F^- diagram (Fig. 53(b)) zero.

If last home wins, or last home loses, we have the disjunctive compound as discussed in the rest of the book. It happens that the games are not tame, so that we have no general theory for the last home loses case. The reader who wishes to use the corresponding table will therefore need to have understood the appropriate parts of Chapter 12.

If last home wins, we move so that the Nim-sum of the numbers in the G^+ diagram (Fig. 54(a)) is zero.

Fig. 54(b) gives a partial strategy for the game when the last home loses. See Chapter 12 for an explanation of the ideas involved.

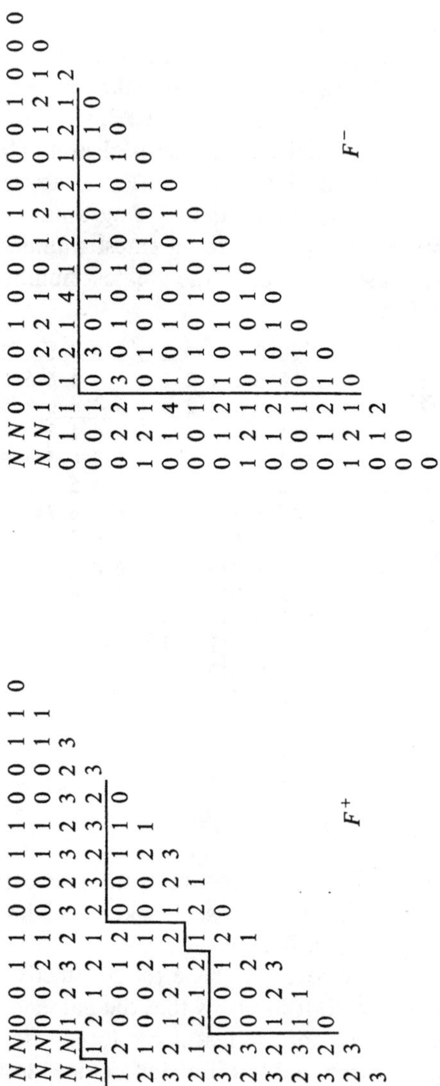

```
P P N N 0 0 1 1 0 0 1 1 0 0 1 1 0
P P N N 0 0 2 1 0 0 1 1 0 0 1 1
N N N N 1 2 3 2 3 2 3 2 3 2 3
N N N 1 2 1 2 1 2 3 2 3 2 3
0 0 1 2 0 0 1 2 0 0 1 1 0
0 0 2 1 0 0 2 1 0 0 2 1
1 2 3 2 1 2 1 2 1 2 3
1 1 2 1 2 1 2 1 2 1
0 0 3 2 0 0 1 2 0
0 0 2 3 0 0 2 1
1 1 3 2 1 2 3
1 1 2 3 1 1
0 0 3 2 0
0 0 2 3
1 1 3
1 1
0
```

F^+

(a)

```
N N 0 0 0 1 0 0 0 1 0 0 0 1 0 0 0
N N 1 0 2 2 1 0 1 2 1 0 1 2 1 0
0 1 1 1 2 1 4 1 2 1 2 1 2 1 2
0 0 1 0 3 0 1 0 1 0 1 0 1 0
0 2 2 3 0 1 0 1 0 1 0 1 0
1 2 1 0 1 0 1 0 1 0 1 0
0 1 4 1 0 1 0 1 0 1 0
0 0 1 0 1 0 1 0 1 0
0 1 2 1 0 1 0 1 0
1 2 1 0 1 0 1 0
0 1 2 1 0 1 0
0 0 1 0 1 0
0 1 2 1 0
1 2 1 0
0 1 2
0 0
0
```

F^-

(b)

FIG. 53.

0	0	1	1	0	0	1	1	0	0	1	1	0	0
0	0	2	1	0	0	1	1	0	0	1	1	0	0
1	2	2	2	3	2	2	2	3	2	2	2	3	
1	1	2	1	4	3	2	3	3	3	2	3		
0	0	3	4	0	0	1	1	0	0	1	1		
0	0	2	3	0	0	2	1	0	0	1			
1	1	2	2	1	2	2	2	3	2				
1	1	2	3	1	1	2	1	4					
0	0	3	3	0	0	3	4						
0	0	2	3	0	0	2							
1	1	2	2	1	1								
1	1	2	3	1									
0	0	3	3										
0	0	2											
1	1												
1													

G^+

(a)

0	0	1	1	0^{0a}	0	1	1	0^{0d}	0	1	1	0^{0j}
0	0	2	1	0	0	1^{1b}	1	0	0	1^{1i}	1	
1	2	2	2	3	2^{142c}	2	2^{2f}	3	2^{1340}	2		
1	1	2	1	4	3^{043e}	2	3^{053g}	3				
0^{0a}	0	3	4	0^{0a}	0	1	1^{1}					
0	0	2^{142c}	3^{043e}	0	0^{0h}	2^{162}	1					
1	1^{16}	2	2	1	2^{162}							
1	1	2^{2f}	3^{053g}	1^{1}	1							
0^{0d}	0	3	3									
0	0	2^{1340}										
1	1^{1i}	2										
1	1											
0^{0j}												

$G*$

(b)

$a = 2_+$, $b = 2_+3 = a + 1$, $c = 2_+1$, $d = b_+$, $e = b420$, $f = ed10$, $g = cb0$, $h = 42$, $i = d3$, $j = i_+$. The games a, b, d, h, i, j are tame.

FIG. 54.

CHAPTER 15

Ups, Downs, and Bynumbers

"Play up! Play up!
and play the game!"
Sir Henry John Newbolt
Vitaï Lampada.

In this chapter, we return to our main theme of partizan games. In particular, we wish to study various natural functions from games to games. We introduce one important kind of function by discussing the theory of a still more general form of Hackenbush.

HACKENBUSH HOTCHPOTCH

We have already met Hackenbush restrained (in Chapter 8), and unrestrained (in Chapter 13). In the restrained form, each edge could be chopped only by its owner, and in the unrestrained form, each edge could be chopped by either player.

In *Hackenbush Hotchpotch*, there are three kinds of edges, *black* (▌) which may only be chopped by Left, *white* (▯), which may only be chopped by Right, and *plain* (|), which may be chopped by either. The rules are otherwise as in the earlier games, which are particular cases. Moreover, the value of any position which has only plain edges may be found by applying the theory of Chapter 13, and the value of many positions with no plain edges by applying the ideas of Chapter 8.

Other values can be found in the usual way. So for instance, we have

$$\triangle = \{ /, \backslash \mid / \} = \{1, * \mid 1\} = 1 \mid 1 = 1 + *.$$

Now seems the appropriate moment to introduce an abbreviating convention for such sums—if x is a number, and $*n$ a Nim-heap, we write $x*n$ for the sum $x + *n$. In the case $n = 1$, we abbreviate $*n$ to $*$, and so we write $x*$ for

$x + *$. This convention will be extended later in the chapter. So we have the equation $\wedge = 1*$.

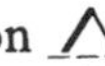

In a similar way, we find

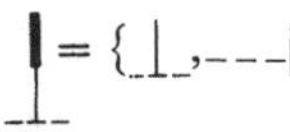

$$= \{\ ,\ \} = \{*, 0 | 0\} = \uparrow + * = \uparrow *, \text{ say.}$$

(The equation $\{*, 0 | 0\} = \uparrow + *$ was not supposed to be obvious—it can be verified by computing the simplest form of $\uparrow + *$.) The abbreviations we here introduce are meant to apply to particular games only, so that in general $2G$ will still mean $G + G$, rather than $2 + G$.

There is usually no risk of confusion between these two meanings (very few people would interpret $2\frac{1}{2}$ as a synonym for 1), but when there is, we write $2.G$ for the meaning $G + G$. Thus $2.\frac{1}{2}$ does mean 1, whereas $2\frac{1}{2}$ means $2 + \frac{1}{2}$. Our convention should be regarded as extending the usual notation for fractions, and we shall observe similar rules.

Now suppose we attach a picture of value x (not necessarily a number) at some point of a Hotchpotch picture P, obtaining a picture $P(x)$, say. How does the value of $P(x)$ depend on x? In Fig. 55 we define three functions $f:x$, $g:x$, $h:x$ in this way.

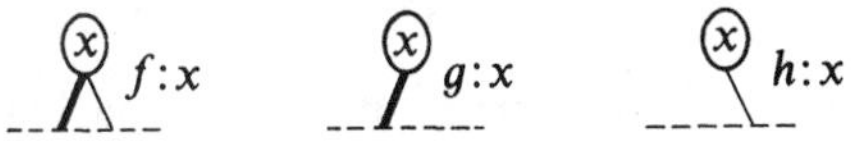

$f:x$ $\quad$ $g:x$ $\quad$ $h:x$

FIG. 55. Some Hotchpotch functions.

Let us see how to define the same functions arithmetically. We have the equation

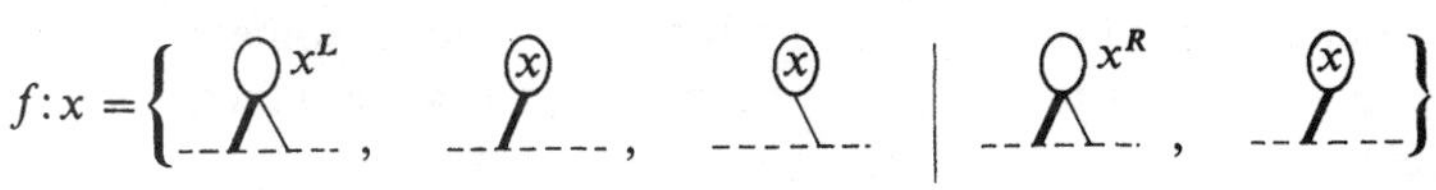

$$f:x = \{\ x^L,\ ,\ \ |\ x^R,\ \}$$

and so $f:x$ is the function defined inductively in terms of the simpler functions $g:x$ and $h:x$ by the formula

$$f:x = \{f:x^L, g:x, h:x \,|\, f:x^R, g:x\}.$$

In a similar way, $g:x = \{g:x^L, 0 \,|\, g:x^R\}$, and our remaining function $h:x$ is defined by $h:x = \{h:x^L, 0 \,|\, h:x^R, 0\}$. How do these functions vary with x?

We call a function $f:x$ a *wop* function (weakly order-preserving) if we have the implication $x \leqslant y$ implies $f:x \leqslant f:y$, and a *sop* (strictly order-preserving) function if $x \leqslant y$ happens if and only if $f:x \leqslant f:y$.

THEOREM 87. *Let there be given any number of wop functions $f_L:x$ and*

$f_R{:}x$, and define a new function $f{:}x$ by

$$f{:}x = \{f{:}x^L, f_L{:}x \mid f{:}x^R, f_R{:}x\}.$$

Then $f{:}x$ is a sop function.

Proof. We play the game $f{:}x - f{:}y$. It is easy to see that any player who can win $x - y$ can win $f{:}x - f{:}y$ in the same circumstances.

As a matter of notation, we write $f{:}x = \{f_L{:} \mid f_R{:}\}{:}x$ for a function defined in this way. If G is a game, we also use G for the function which identically takes the value G, and omit the colon.

Thus our examples were $f{:}x = \{g{:}, \ h{:} \mid g{:}\}{:}x$, $g{:}x = \{0 \mid \}{:}x$, and $h{:}x = \{0 \mid 0\}{:}x$. Since the identically zero function is wop, the functions $f{:}, g{:}, h{:}$ are all sop.

Notice that the value of a wop or sop function $f{:}x$ depends only on the *value* of x and not on its *form*, for if $x = y$, then $x \geqslant y$ and $x \leqslant y$, so $f{:}x \geqslant f{:}y$ and $f{:}x \leqslant f{:}y$, so that $f{:}x = f{:}y$. Notice also that after the discussion of Hackenbush in Chapter 13, we can regard the functions defined here as generalisations of animating functions.

DIGITAL DELETIONS

This game is played with a string of decimal digits, perhaps the string

8315553613086720000.

It is an impartial game, and the player to move may *either* DECREASE any digit, leaving the others unaltered, or DELETE any digit 0 and all following digits of the string.

It follows from the preceding theory that if we precede some string of value x with a digit n, the value of the resulting string is a sop function $f_n{:}x$. For instance

$$f_3{:}x = \{f_3{:}x', f_2{:}x, f_1{:}x, f_0{:}x\}. \quad f_0{:}x = \{f_0{:}x', 0\}.$$

are the inductive definitions of f_3 and f_0. Since the values of impartial games are Nim-heaps $*N$, we need only tabulate the functions f_n (Fig. 56).

In this table, we have written $f_n{:}x = y$, when we really mean that $f_n{:}*x = *y$. Now let us see what move to make from the string 314159. This string has Grundy number $f_3{:}(f_1{:}(f_4{:}(f_1{:}(f_5{:}(f_9{:}0)))))$, since it is followed by the empty sequence, whose Grundy number is 0.

We evaluate this as 12:

$$12 \xleftarrow[f_3]{} 10 \xleftarrow[f_1]{} 10 \xleftarrow[f_4]{} 7 \xleftarrow[f_1]{} 7 \xleftarrow[f_5]{} 9 \xleftarrow[f_9]{} 0$$

n	0	1	2	3	4	5	6	7	8	9	10	11	12	13	14	15	16	17	18	19	20	21
$f_0:n$	1	2	3	4	5	6	7	8	9	10	11	12	13	14	15	16	17	18	19	20	21	22
$f_1:n$	0	1	2	3	4	5	6	7	8	9	10	11	12	13	14	15	16	17	18	19	20	21
$f_2:n$	2	0	1	5	3	4	8	6	7	11	9	10	14	12	13	17	15	16	20	18	19	23
$f_3:n$	3	4	0	1	2	7	5	9	6	8	12	13	10	11	16	14	18	15	17	21	22	19
$f_4:n$	4	3	5	0	1	2	9	10	11	6	7	8	15	16	12	13	14	19	21	17	18	20
$f_5:n$	5	6	4	2	0	1	3	11	10	7	8	9	16	15	17	12	13	14	22	23	24	18
$f_6:n$	6	5	7	8	9	0	1	2	3	4	13	14	11	10	18	19	12	20	15	16	17	24
$f_7:n$	7	8	6	9	10	3	0	1	2	5	4	15	17	18	11	20	19	12	13	14	16	25
$f_8:n$	8	7	9	6	11	10	2	0	1	3	5	4	18	17	19	21	20	13	12	15	14	16
$f_9:n$	9	10	8	7	6	11	4	3	0	1	2	5	19	20	21	18	22	23	14	12	13	15

FIG. 56.

Now we want to move to some position with Grundy number 0, so we "pretend" the answer is 0, and work backwards:

$$0 \xleftarrow[f_3]{} 2 \xleftarrow[f_1]{} 2 \xleftarrow[f_4]{} 5 \xleftarrow[f_1]{} 5 \xleftarrow[f_5]{} 0 \xleftarrow[f_9]{} 8$$

We can now make our dreams come true by finding what changes in the individual digits enable us to pass from one of these chains to the other. Most of these require digits to be increased, but the last one is legal, since

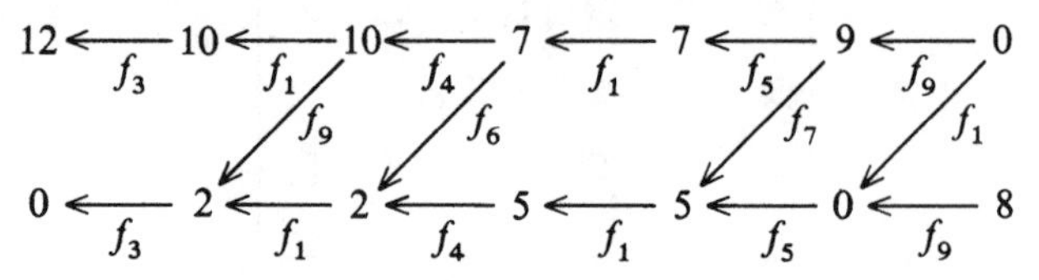

it requires us to replace f_9 by f_1. So the only winning move here is to move to 314151. The reader might like to verify that the position

8315553613086720000

yields just two winning moves—to decrease 7 to 6, or delete the last two zeros.

The inductive definitions of the f_n tell us that each entry of the table is the mex of the numbers above and to the left of it, except that 0 is not allowed as an entry in the f_0 line. We can deduce that the entries in each line are ultimately arithmetico-periodic, so that the game has in principle a complete theory. Perhaps some reader will find out exactly when the periodicity occurs. But apart from the formulae $f_0{:}x = x + 1$, $f_1{:}x = x$, $f_2{:}x = x +_3 2$, and $f_3{:}(x + 9) = f_3{:}x + 9$ for $x \geqslant 3$, there seem to be no easy answers.

ORDINAL ADDITION OF GAMES

A particularly important case of these inductively defined functions is when each of the f_L and f_R is a constant function. So the function

$$f{:}x = \{A, B, C, \ldots{:}\,|\, D, E, F, \ldots\}{:}x$$

(when $A, B, C, \ldots, D, E, F, \ldots.$ are any games) is inductively defined by

$$f{:}x = \{A, B, C, \ldots, f{:}x^L \,|\, D, E, F, \ldots, f{:}x^R\}.$$

In other words, we obtain the tree of the game $\{A, B, C, \ldots | D, E, F, \ldots\}{:}x$ by adding new moves for Left to $A, B, C, \ldots$ and for Right to $D, E, F, \ldots$ *at every position* of the game x.

It follows from the previous discussion that $\{A, B, C, \ldots | D, E, F, \ldots\}{:}x$ is a sop function of x, and so in particular, that it depends only on the value of the game x, and not on its form. We can also regard it as a function of the game $\{A, B, C, \ldots \mid D, E, F, \ldots\}$, but it then depends in part on the form of this game, and not only its value.

For instance, consider the games $0 = \{\,\mid\,\}$, and $\varphi = \{-1 \mid 1\}$. These games have equal value. But the games $\{\,\mid\,\}:1$ and $\{-1 \mid 1\}:1$ do not. In fact, $\{\,\mid\,\}:1$ is the game obtained from 1 by adding no new moves at any position—namely 1, whereas $\{-1 \mid 1\}:1$ is obtained by adding new moves (to -1 for Left, 1 for Right) at each node of the tree for 1, giving the game

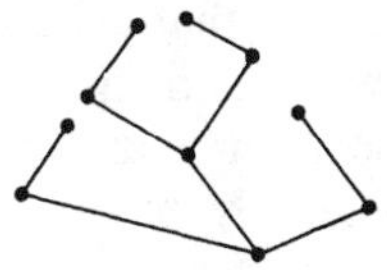

$$= \{-1, -1 \mid 1 \parallel 1\} = \{-1, 0 \mid 1\} = \tfrac{1}{2}.$$

However, if the game $G = \{A, B, C, \ldots \mid D, E, F, \ldots\}$ is supposed to have no reversible moves, we can define a function depending only on the value of G by

$$G{:}X = \{A, B, C, \ldots \mid D, E, F, \ldots\}{:}\,X,$$

since it turns out that if also H has no reversible moves, and $H = G$, then $H{:}X = G{:}X$. Since every game G has a form without reversible moves, the function $G{:}X$ is defined for all G and X, and we call it the *ordinal sum* of G and X.

As an example, we consider the ordinal sum $\frac{1}{2}{:}x$, and restrict ourselves at first to the case when x is a number. Since $\frac{1}{2} = \{0 \mid 1\}$ without reversible moves, $\frac{1}{2}{:}x$ is defined by adding new options 0 for Left and 1 for Right at every position of x. In other words, we tell the creation story over again, but examine only the space between 0 and 1.

So for instance the simplest number born here is $\frac{1}{2}$, so $\frac{1}{2}{:}0 = \frac{1}{2}$. The simplest numbers to the left and right of this (but still between 0 and 1) are $\frac{1}{4}$ and $\frac{3}{4}$, so $\frac{1}{2}{:}-1 = \frac{1}{4}$, $\frac{1}{2}{:}1 = \frac{3}{4}$. Similarly, we find that for $x = \frac{1}{2}, 2, 3, \ldots, \omega$, the values of $\frac{1}{2}{:}x$ are $\frac{5}{8}, \frac{7}{8}, \frac{15}{16}, \ldots, 1 - 1/\omega$.

In general, to see what $x{:}y$ means when x and y are both numbers, we refer to the tree in chapter 0, or equivalently, to the so-called sign-expansions used in chapter 3. We get from 0 to $x{:}y$ in the tree by starting at the point x and making exactly those moves along the tree which would get us from 0 to y. For example, we get from 0 to $\frac{3}{4}$ by moving right, left, right, and so we get from $\frac{1}{2}$ to $\frac{1}{2}{:}\frac{3}{4}$ by moving right, left, right again, arriving at $\frac{11}{16}$. Similarly, ω steps leftward from $\frac{1}{2}$ get us to $1/\omega$, and so $\frac{1}{2}{:}-\omega = 1/\omega$. In general, the sign-expansion of $x{:}y$ is that of x followed by that of y.

When x is a number and G is an arbitrary game, we can compute $x{:}G$ as follows. Play G until we get to its *stopping positions* (Chapter 9). In the tree of G, replace each stopping position y by the ordinal sum $x{:}y$. So for example $\frac{1}{2}{:}\pm 1 = \{\frac{1}{2}{:}1 \mid \frac{1}{2}{:}-1\} = \frac{3}{4} \mid \frac{1}{4}$.

When X is not a number, the behaviour of $X:Y$ is more complicated. As a general rule, we can say that we expect its value to be "very near" that of X, and this expectation is given a more formal meaning by Simon Norton's theorem that X and $X:Y$ have the same order-relations with all games G not involving X (as a position). We shall give Norton's proof of this in the next chapter, and then deduce, following Norton, that $X:Y$ is free from reversible moves if X and Y are, and so that we have the associative law $(X:Y):Z = X:(Y:Z)$.

The particular case $X = *$ gives rise to a family of games that arise very frequently. The same games can be defined in many ways, and the order-relations in the group they generate together with the Nim-heaps $*n$ are completely known. For positive numbers x, and more generally for games all of whose stopping positions are positive numbers, we define the game $\uparrow x$ (pronounced "up-x") by the formula

$$* + \uparrow x = \{0 \mid 0\}:x = *:x.$$

It turns out that the same games can be defined by the formulae

$$\uparrow x = \{\ \mid *\}:x = \{* \mid *\}:x$$

under the same conditions on x. The negative of $\uparrow x$ is called $\downarrow x$ (down-x). The following theorem, which we do not prove, gives the complete structure of the group generated by the $\uparrow x$ and the $*n$, for numbers x.

THEOREM 88. *Let X be a finite sum of terms $\uparrow x$ and $\downarrow y$, in which all the numbers x and y are positive, and no number x occurs also as a y (for then we could cancel). Then X is positive if and only if either the number of $\uparrow x$ terms exceeds the number of $\downarrow y$ terms, or these numbers are equal, and the least of the numbers x and y is a y. The game $X + *$ is positive if and only if $X + \uparrow \mathbf{On}$ is positive, where* **On** *temporarily denotes any number bigger than all the x and y. The game $X + *n$, with $n \geqslant 2$, is positive only if the number of $\uparrow x$ terms exceeds the number of $\downarrow y$ terms.*

It is often convenient, in seeing what this theorem tells us about the size of the game $\uparrow x$, to use the following symbolic formula:

$$\uparrow x = \uparrow \mathbf{On}(1 - \uparrow^x),$$

where $\uparrow^x$ denotes "the xth power of $\uparrow$". If we suppose, as is natural, that whenever $x > y \geqslant 0$, the xth power of $\uparrow$ is infinitesimal with respect to the yth power, and suppose $\uparrow x < \uparrow \mathbf{On}$ for all x, this formula gives us the right order relations. Note that when we put $x = 1$, it yields the symbolic formula

$$\uparrow \mathbf{On} = \frac{\uparrow}{1 - \uparrow} = \uparrow + \uparrow^2 + \uparrow^3 + \ldots$$

(the infinite series on the right side of this, is, like the whole equation, purely formal, and should best be thought of as extending over all ordinal numbers α).

So the games $\uparrow x$, for $1 \leqslant x < \mathbf{On}$, are very close to $\uparrow$, being between $\uparrow$ and $\uparrow + \uparrow^2 + \uparrow^3 + \dots$. For integral x, they are the partial sums of this series, thus

$$\uparrow 2 = \uparrow + \uparrow^2, \quad \uparrow 3 = \uparrow + \uparrow^2 + \uparrow^3, \dots$$

The game $\uparrow^2$ can be defined as $\{0|\downarrow *\}$, and $\uparrow^3$ as $\{0|\downarrow 2*\}$, etc. So that we can write $\downarrow_2, \downarrow_3, \dots$ for their negatives, we pronounce these games "up-second", "up-third", etc., and the negatives similarly "down-second", "down-third", and so on.

If we adopt these conventions, Theorem 88 tells us that the game $*$ is confused with all sums between $\downarrow\mathbf{On}$ and $\uparrow\mathbf{On}$, greater than all smaller sums, and less than all larger ones. Note that we need not enquire about the critical values $\downarrow\mathbf{On}$ and $\uparrow\mathbf{On}$ themselves, since these are not real games, but purely formal symbols. The theorem also tells us that for $n \geqslant 2$, $*n$ is confused with all sums with as many up terms as down ones, but not with any other sums—an elegant way of putting this is that such a $*n$ is confused with a sum X if and only if $*$ is confused with $X + X$.

SHRINKING RECTANGLES

In this game, played with a number of rectangles of integer sides, Left may decrease the breadth of any rectangle, and Right the height. A rectangle whose breadth or height is decreased to zero disappears. What are the values?

Here since either player may shrink his coordinate to zero, the moves to 0 are always available from every position in the game corresponding to a single rectangle. So such positions have the form $*:x$ for some x, and when we try to tabulate them, it is obvious that a rectangle of breadth b and height h has value $*:(b-h)$ which in our standard notation is $* + \uparrow(b-h)$. So Theorem 88 can be used to give a complete analysis of this game.

(*Note*. We define $\uparrow(-x) = -\uparrow x = \downarrow x$ for all x for which $\uparrow x$ is defined.)

THE GAMES $\Uparrow x$, $(3.\uparrow)x$

It is sometimes convenient to define

$$\Uparrow x = \{|\uparrow *\}:x \quad (3.\uparrow)x = \{|\Uparrow *\}:x,$$

and so on. The abbreviations $\uparrow *$, $\Uparrow *$, ... mean $\uparrow + *$, $\uparrow + \uparrow + *$, ..., and $\Uparrow *$, for instance, is pronounced "double-up-star". These definitions are equivalent to

$$\Uparrow x* = \{|\uparrow\}:x, \qquad (3.\uparrow)x* = \{|\Uparrow\}:x, \dots$$

for all x whose stopping positions are all positive. When x is a *number*, we have $⇑x = ↑ + ↑x$, $(3.↑)x = ↑ + ↑ + ↑x$, etc., provided $x \geqslant 1$, while for numbers x with $0 < x < 1$, $⇑x$ behaves like $↑ + ↑x^+$, where x^+ behaves as a number less than 1, but greater than all such numbers which appear in terms $↑y$. Similarly $(3.↑)x$ behaves like a sum $↑ + ↑ + ↑x^{++}$, where $x^{++} = x$ for $x \geqslant 1$, but we have $y^+ < x^{++} < 1$ for all y^+, if $x < 1$, and so on. These observations enable us to extend Theorem 88 to cover sums and differences of terms $⇑x$, $(3.↑)x$, etc.

Thus

$$⇑ + ↑\tfrac{1}{2} < ⇑ + ↑\tfrac{3}{4} < ↑ + ⇑\tfrac{1}{4} < ↑ + ⇑\tfrac{1}{2} < (3.↑)\tfrac{1}{2} < 3.↑ = ↑ + ↑ + ↑.$$

In general, sums of these games with each other, with terms $*n$, and with numbers, are written by juxtaposition, thus

$$3\tfrac{1}{2}↑2↓3⇑\tfrac{1}{16}*5 \text{ means } 3 + \tfrac{1}{2} + ↑2 + ↓3 + ⇑\tfrac{1}{16} + *5.$$

$↑1$, $⇑1$, $(3.↑)1$ are written simply $↑$, $⇑$, $3.↑$, and represent the sums

$$↑, ↑ + ↑, ↑ + ↑ + ↑.$$

More generally, we use the symbol $\hat{n}$ for $n.↑ = ↑ + ↑ + \ldots + ↑$, and $\hat{n}x$ with the obvious meaning extending that of $↑x$, $⇑x$, $(3.↑)x$.

To see how these games arise in "real life", we consider yet another example.

THE TROMINO GAME

This game is played in a finite strip of squares. Left has an infinite supply of black straight trominoes (i.e. 3×1 rectangles), and Right an infinite supply of white ones. Initially, a black tromino is placed at one end of the strip, and a white one at the other. Then the two players play alternately, each placing one of his trominoes somewhere in the strip (so as exactly to cover three empty squares) subject to the condition that trominoes of the same colour may not touch.

The values can be worked out completely, with some patience. (Richard K. Guy and I once had such patience!) It turns out that they have a curious kind of arithmetico–geometric ultimate periodicity with period 13, increment $*$, and multiplier $\frac{1}{4}$! Figure 57 gives the values—we tabulate for each n, the value of a strip of n squares bounded at both ends by white trominoes (the value when both ends are bounded by black trominoes being the negative of this), and the value when the ends are bounded by trominoes of opposite colours. There is an easy argument which shows the latter kind of value must always have the form $*n$.

Ultimately, when n is increased by 13, $*$ is added to each entry, and the argument of each $⇑$ is multiplied by $\frac{1}{4}$. The arrows indicate exceptions to the ultimate behaviour, which all affect the *ww* column only.

	ww	*wb*		*ww*	*wb*		*ww*	*wb*
→ 0	—	0	13	$\uparrow *$	$*$	26	$\uparrow$	0
→ 1	0	0	→14	$*$	$*$	27	$\uparrow$	0
2	0	0	15	$*$	$*$	28	0	0
→ 3	1	0	16	$\Uparrow *$	$*$	29	$\Uparrow\frac{1}{4}$	0
→ 4	1	$*$	17	$\Uparrow\frac{1}{2}$	0	30	$\Uparrow\frac{1}{8}*$	$*$
5	$*$	$*$	18	0	0	31	$*$	$*$
→ 6	$*$	$*$	19	$\uparrow$	0	32	$\uparrow *$	$*$
7	$\uparrow *$	$*$	20	$\uparrow$	0	33	$\uparrow *$	$*$
8	$\frac{1}{2}$	$*2$	21	$\Uparrow\frac{1}{2}*$	$*3$	34	$\Uparrow\frac{1}{8}$	$*2$
9	$\uparrow$	0	22	$\uparrow *$	$*$	35	$\uparrow$	0
→10	0	0	23	$\uparrow *$	$*$	36	$\uparrow$	0
11	$\uparrow$	0	24	$\uparrow *$	$*$	37	$\uparrow$	0
12	$\Uparrow$	$*3$	25	$\Uparrow\frac{1}{4}*$	$*2$	38	$\Uparrow\frac{1}{16}$	$*3$

FIG. 57. Values in the tromino game.

THE TROMINO GAME WITH FREE ENDS. CIRCULAR TROMINOES

Berlekamp has extended the analysis to cover the case in which the strip of squares need not start with any terminal trominoes, so that either player can move at the end. He finds that there is a similar type of arithmetico–geometric ultimate periodicity, and observes that some new values occur. Fig. 58 gives his results (*o* denoting a free end), and we add a further column showing the values when the initial configuration consists of a strip of *n*

	ow	*oo*	circle		*ow*	*oo*	circle		*ow*	*oo*	circle
0	0	0	0	13	$*$	$*$	$*$	26	0	0	0
1	0	0	0	14	$*$	$*$	0	27	0	0	0
2	0	0	0	15	$*$	$*$	0	28	0	0	0
3	1	$*$	0	16	$\Uparrow\frac{1}{2};1$	0	0	29	$\Uparrow\frac{1}{4};\frac{1}{8}$	$*$	0
4	$*$	$*$	$*$	17	0	0	0	30	$*$	$*$	0
5	$*$	$*$	$*$	18	0	0	0	31	$*$	$*$	$*$
6	$*$	$*$	0	19	0	0	0	32	$*$	$*$	0
7	$\uparrow *$	$*2$	0	20	$\uparrow\frac{1}{4}$	$*2$	0	33	$\uparrow\frac{1}{16}*$	$*2$	0
8	$\uparrow$	0	0	21	$\uparrow *$	$*$	$*$	34	$\uparrow$	0	0
9	0	0	0	22	$*$	$*$	0	35	0	0	0
10	0	0	0	23	$*$	$*$	0	36	0	0	0
11	$\uparrow$	$*2$	0	24	$\uparrow *$	$*2$	0	37	$\uparrow$	$*2$	0
12	$\uparrow\frac{1}{2}*$	$*$	0	25	$\uparrow\frac{1}{8}$	0	0	38	$\uparrow\frac{1}{32}*$	$*$	0

FIG. 58.

squares bent round into a circle, which are easily deduced from Berlekamp's results.

Here we see the values $\uparrow\frac{1}{2}$, $\uparrow\frac{1}{4}$, ..., which we explained a few pages ago, and also a new system of values of the form $\Uparrow x;y$. These new games are defined by the formula

$$\Uparrow x;y = \{0 \mid \uparrow x*, \uparrow y\}.$$

I have observed them occurring also in other games, for a wide range of values of x and y. Notice that the periodicity of the last two columns is exact, with period 26.

THE SUBMULTIPLES OF $\uparrow$

It is natural to ask whether there is any game G with $G + G = \uparrow$? If so, then of course there will be many such games, for from any one we can obtain another by adding some game H of order 2. But in fact we can prove that any game G can be halved.

THEOREM 89. *For all sufficiently large n, if we define* $H = \{n \mid G - n\}$, *then* $H + H = G$.

[For long games G, n may need to be some ordinal.]

Proof. Play the game $H + H - G$. Each of the moves in the H components has its counter, and if n is sufficiently large, each move in G will be countered by some move in H.

Notice that this applies even when $G = 0$, and gives us an infinity of distinct games of order 2, which can be halved to give us games of orders 4, 8, 16, ... On the other hand, a fairly long-standing conjecture, recently proved by Norton, asserts that there is no short game of any *odd* order. (Norton has also found long games of *all* finite orders.)

We can modify the argument to produce what appear to be the simplest submultiples of $\uparrow$. In general, we define a game $\hat{x} = x.\uparrow$ as follows.

If x is a positive integer (or zero), $\hat{x}$ is the sum of x copies of $\uparrow$. If x is a negative integer, $\hat{x}$ is the sum of $-x$ copies of $\downarrow$. Otherwise, if $x = \{x^L | x^R\}$, we define $\hat{x}$ by the formula

$$\hat{x} = \{x^L + \Uparrow * \mid x^R + \Downarrow *\}.$$

It turns out that then we also have

$$\hat{x}* = \{x^L + \Uparrow \mid x^R + \Downarrow\}.$$

The definition is invalid when we take $x = \omega$, and indeed it does not seem possible to define $\hat{\omega}$ in any natural way. But it works quite happily at, for instance $x = \frac{1}{3}$ or $x = 1 + 1/\omega$.

According to this definition, we have, for example

$$\hat{\tfrac{1}{2}} = \{\Uparrow * \mid \downarrow *\}, \qquad \hat{\tfrac{1}{2}}* = \{\Uparrow \mid \downarrow\}$$
$$\hat{\tfrac{3}{2}} = \{(3.\uparrow)* \mid *\}, \qquad \hat{\tfrac{3}{2}}* = \{3.\uparrow \mid 0\}.$$

The games $\hat{\frac{1}{2}}$ and $\hat{\frac{3}{2}}$ are pronounced "semi-up" and "sesqui-up".

The definition has the properties we should hope for, namely that $\widehat{x+y} = \hat{x} + \hat{y}$, $\widehat{-x} = -\hat{x}$, etc. Moreover, these games can actually arise in "real-life" games.

BYNUM'S GAME, AND TWISTED BYNUM'S GAME

The following game was suggested by Jim Bynum. Start with a rectangular array of cards: later on in the game this will become a number of separated arrays. Left moves by removing a vertical strip from just one of these rectangles, and Right by removing a horizontal strip. The strip's length must be the full height or breadth of the rectangle respectively, so that it splits the rectangle into two, unless it is removed from the edge, when it leaves a slightly smaller rectangle.

We shall give an analysis of Bynum's game at the end of the chapter, since it seems to be one of the most interesting games we have studied. It is fairly easy to see what the outcome is from any initial position involving

	1	2	3	4	5	6	7	8	9	10	11	12	13	14	15
1	$*$	$\uparrow$	$*$	$\uparrow^2$	$\downarrow*$	$\uparrow^2$	$*$	$\uparrow$	$*$	$\uparrow^2$	$\downarrow*$	$\uparrow^2$	$*$	$\uparrow$	$*$
2	$\downarrow$	0	$\downarrow$	0	$\downarrow_2$	0	$\downarrow$	0	$\downarrow$	0	$\downarrow_2$	0	$\downarrow$	0	$\downarrow$
3	$*$	$\uparrow$	$*$	$\uparrow^2$	$\hat{*}*$	$\uparrow^2$	$*$	$\uparrow$	$*$	$\uparrow^2$	$\hat{*}*$	$\uparrow^2$	$*$	$\uparrow$	$*$
4	$\downarrow_2$	0	$\downarrow_2$	0	$\downarrow_3$	0	$\downarrow_2$	0	$\downarrow_2$	0	$\downarrow_3$	0	$\downarrow_2$	0	$\downarrow_2$
5	$\uparrow*$	$\uparrow^2$	$\hat{*}*$	$\uparrow^3$	$*$	$\uparrow^2$	$\hat{\frac{1}{2}}*$	$\uparrow^3$	$\hat{\frac{1}{4}}*$	$\uparrow^2$	$*$	$\uparrow^3$	$\hat{\frac{1}{2}}*$	$\uparrow^2$	$\hat{\frac{1}{4}}*$
6	$\downarrow_2$	0	$\downarrow_2$	0	$\downarrow_2$	0	$\downarrow_2$	0	$\downarrow_2$	0	$\downarrow_2$	0	$\downarrow_2$	0	$\downarrow_2$
7	$*$	$\uparrow$	$*$	$\uparrow^2$	$\check{\frac{1}{2}}*$	$\uparrow^2$	$*$	$\uparrow$	$*$	$\uparrow^2$	$\check{\frac{1}{2}}*$	$\uparrow^2$	$*$	$\uparrow$	$*$
8	$\downarrow$	0	$\downarrow$	0	$\downarrow_3$	0	$\downarrow$	0	$\downarrow$	0	$\downarrow_3$	0	$\downarrow$	0	$\downarrow$
9	$*$	$\uparrow$	$*$	$\uparrow^2$	$\check{\frac{1}{4}}*$	$\uparrow^2$	$*$	$\uparrow$	$*$	$\uparrow^2$	$\check{\frac{1}{4}}*$	$\uparrow^2$	$*$	$\uparrow$	$*$
10	$\downarrow_2$	0	$\downarrow_2$	0	$\downarrow_2$	0	$\downarrow_2$	0	$\downarrow_2$	0	$\downarrow_2$	0	$\downarrow_2$	0	$\downarrow_2$
11	$\uparrow*$	$\uparrow^2$	$\hat{*}*$	$\uparrow^3$	$*$	$\uparrow^2$	$\hat{\frac{1}{2}}*$	$\uparrow^3$	$\hat{\frac{1}{4}}*$	$\uparrow^2$	$*$	$\uparrow^3$	$\hat{\frac{1}{2}}*$	$\uparrow^2$	$\hat{\frac{1}{4}}*$
12	$\uparrow^2$	0	$\uparrow^2$	0	$\uparrow^3$	0	$\uparrow^2$	0	$\uparrow^2$	0	$\uparrow^3$	0	$\uparrow^2$	0	$\uparrow^2$
13	$*$	$\uparrow$	$*$	$\uparrow^2$	$\check{\frac{1}{2}}*$	$\uparrow^2$	$*$	$\uparrow$	$*$	$\uparrow^2$	$\check{\frac{1}{2}}*$	$\uparrow^2$	$*$	$\uparrow$	$*$
14	$\downarrow$	0	$\downarrow$	0	$\downarrow_2$	0	$\downarrow$	0	$\downarrow$	0	$\downarrow_2$	0	$\downarrow$	0	$\downarrow$
15	$*$	$\uparrow$	$*$	$\uparrow^2$	$\check{\frac{1}{4}}*$	$\uparrow^2$	$*$	$\uparrow$	$*$	$\uparrow^2$	$\check{\frac{1}{4}}*$	$\uparrow^2$	$*$	$\uparrow$	$*$

FIG. 59.

just one rectangle, but from arbitrary positions the theory is incomplete, but will handle any position likely to be met with in practice.

In the twisted form of Bynum's game, before making any move, the player must rotate the rectangle through one right angle. Otherwise the rules are as before. When we tabulate the values (Fig. 59) the table exhibits a periodicity which can fairly easily be proved to persist, and so we have a complete theory.

Values for the twisted form of Bynum's game.

In each row and column, the twelve entries after the first three recur indefinitely. $\hat{*}*$ means $\pm\Uparrow$, and $\check{\frac{1}{2}}, \check{\frac{1}{4}}$ are the negatives of $\hat{\frac{1}{2}}, \hat{\frac{1}{4}}$.

Notice that the values $\uparrow^2, \uparrow^3$, and $\pm\Uparrow$ appear, as well as certain submultiples of $\uparrow$. To see who wins any given position, proceed as follows. If there is any term $\pm\Uparrow$, this is the best move for the first player—if not, a term $\hat{\frac{1}{4}}*$ or $\check{\frac{1}{4}}*$ will be the best move, or failing that, $\hat{\frac{1}{2}}*$ or $\check{\frac{1}{2}}*$. When all these moves have been made, the value is in the group considered in Theorem 88, and so the winner is known.

It is remarkable that a game with such a simple definition can at one and the same time have a complete theory, yet such a complicated one in play. The peculiar emphasis placed on numbers $6m + 5$ is surprising.

CUTCAKE

We digress for a moment to consider a game which is defined in a similar way to Bynum's game, but which turns out to have a very simple theory.

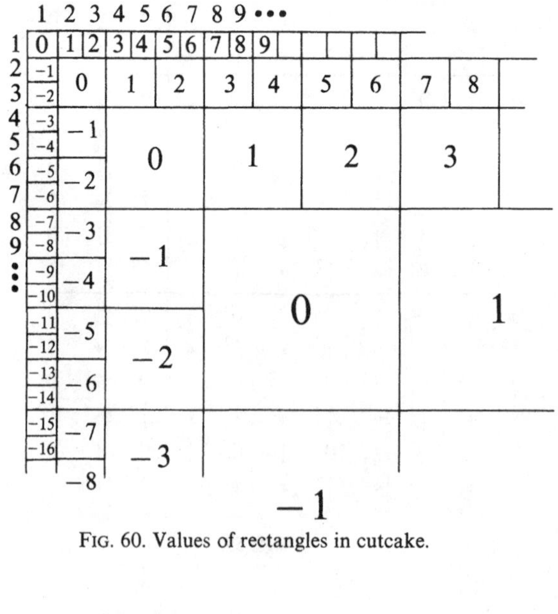

FIG. 60. Values of rectangles in cutcake.

The game is played with a number of rectangular pieces of thin cake, already scored into squares by horizontal and vertical lines. Left moves by breaking some piece into two smaller pieces by breaking it along some horizontal line, and Right moves by breaking some piece along a vertical line. When the cake has been completely broken up into squares of the minimal size, the player who made the last move eats the cake.

Once again, the typical position is a disjunctive sum of positions corresponding to the individual pieces of cake, and so we need only tabulate values at these. Obviously pieces which are long in the horizontal direction tend to favour Left, since he has more scope for vertical breaks. But we think the complete answer is rather surprising. See Figure 60.
So for instance a 4×7 rectangle still has value 0, although at first sight it might appear to favour the player whose cuts in it are shorter.

THE ANALYSIS OF BYNUM'S GAME

As always, the proper thing to do is to tabulate small positions and their values. When we do this with Bynum's game, certain patterns emerge:

	1	2	3	4	5	6
1	$*$	$\downarrow$	$*$	$\downarrow$	$*$	$\downarrow$
2	$\uparrow$	$*$	X	$X\downarrow *$	Y	$Y\downarrow *$
3	$*$	$-X$	$*$	$-X$	$*$	$-X$
4	$\uparrow$	$X\uparrow *$	X	$*$	Y	

$X = \Uparrow | *$
$Y = X\uparrow | *$

Preliminary analysis of Bynum's game

These suggest the following result, which we call the *Theorem of* 17 *October*:
The values in Bynum's game are given by the following scheme

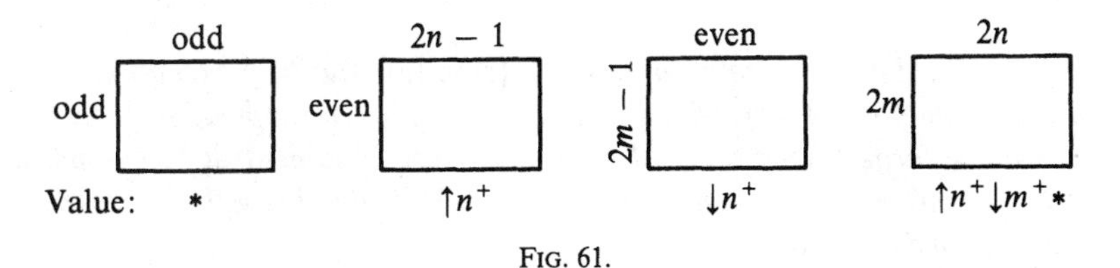

FIG. 61.

where the game $\uparrow 1^+ = \uparrow = 0 | *$, and for larger n, $\uparrow n^+$ is the game, with negative $\downarrow n^+$, defined by the formula

$$\uparrow n^+ = \{\uparrow a^+ + \uparrow b^+ \mid *\}_{a+b=n}.$$

Since $\uparrow n^+$ is positive for all n, we can add 0 as a Left option, if we like, in this definition.

This theorem is equivalent to the assertion that Bynum's game is abstractly equivalent to the game soon to be defined, which we shall call **Bynumbers.** We do not need to prove the theorem, since in play it is very easy to see the correspondence between moves in the two games, or prove that some of the moves are stupid ones. We deduce easily that the initial rectangle in Bynum's game is a win for the first player if both coordinates have the same parity, and otherwise a win for the player whose coordinate is even.

THE GAME OF BYNUMBERS

Each player has a number of heaps of counters, and there may also be some neutral heaps. Your legal moves are:

(a) to neutralise any one of your opponent's heaps,

(b) to split any one of your own heaps into two non-empty parts,

(c) to throw away any neutral heap,

(d) to throw away one of your own heaps.

A heap of size n belonging to Left has value $\uparrow n^+$, so that one of the same size belonging to Right will have value $\downarrow n^+$. Neutral heaps all have the value $*$, their exact size being immaterial. The moves allowed correspond to the definition

$$\uparrow n^+ = \{\uparrow a^+ + \uparrow b^+, 0 \mid *\}_{a+b=n}$$

which is valid for all n.

A great many results have been proved about this game, mostly by P. T. Johnstone and M. R. Christie. We can summarise a number of them in Christie's strategic rules (in which we have supposed that any pairs of neutral heaps have been cancelled, using $* + * = 0$):

One should usually prefer (a) *to* (b) *to* (c) *to* (d). *The only exception is that* (b) *may sometimes be preferred to* (a), *but only when the heap to be split is the unique largest heap on the board, and there is no neutral heap* (*and not always then*). *Moreover, any move of type* (a) *should always be to neutralise one's opponent's largest heap.*

When we say that one should prefer X to Y, we merely mean that if Y is a good move, then so will X be, and not any stronger assertion. The proof of these rules is quite subtle, but they make the play very simple, for in many cases they leave only a unique move to be considered. In particular, it is easy to see that a player who has two or more more heaps than his opponent, or one heap more and the move, has an easy win.

In the further analysis, it is convenient to write

$$\uparrow n^{+} = \uparrow + \uparrow^{2+} + \uparrow^{3+} + \ldots + \uparrow^{n+}.$$

Then the values $\uparrow, \uparrow^{2+}, \uparrow^{3+}, \ldots$ are all positive all small games, whose order relations can be fairly easily investigated using Christie's rules. Any position in Bynum's game is a sum of such values and their negatives, together possibly with $*$.

But Christie's rules show that when $*$ occurs and the first player has any good move, then such a move is to neutralise his opponent's largest heap, or remove the neutral heap when the opponent has no heap. So to analyse the game completely, we need only consider sums and differences of the values $\uparrow^{n+}$ (counting $\uparrow = \uparrow^{1+}$).

A rather surprising consequence of the rules is that if such an expression is positive, it will remain positive however we change the coefficients of those terms $\uparrow^{n+}$ that appear a negative number of times in all. (For in the corresponding game of Bynumbers, coefficients less than or equal to -2 will at the appropriate stage in the game cause the largest heap to be non-unique, and then the players will alternately neutralise each other's largest heaps until uniqueness is restored.) So when testing whether such an expression is positive, we can suppose each negative term appears with coefficient -1 when equal terms are collected and cancelled with occurrences of their negatives.

From this it follows that there are only a finite number of basic inequalities involving numbers up to any given size n, which Christie and I have calculated up to $n = 8$. We have

$$\begin{array}{ll}
\uparrow^{2+} \gg \uparrow^{3+} & \uparrow^{1+} + \uparrow^{1+} \gg \uparrow^{2+} \\
\uparrow^{2+} \gg \uparrow^{4+} & \uparrow^{3+} + \uparrow^{4+} \gg \uparrow^{6+} + \uparrow^{7+} \\
\uparrow^{3+} \gg \uparrow^{5+} & \uparrow^{2+} + \uparrow^{3+} \gg \uparrow^{6+} + \uparrow^{8+} \\
\uparrow^{2+} \gg \uparrow^{6+} & \uparrow^{2+} + \uparrow^{4+} \gg \uparrow^{6+} + \uparrow^{8+} \\
\uparrow^{3+} \gg \uparrow^{7+} & \uparrow^{2+} + \uparrow^{6+} \gg \uparrow^{7+} + \uparrow^{8+} \\
\uparrow^{4+} \gg \uparrow^{7+} & \uparrow^{3+} + \uparrow^{4+} \gg \uparrow^{7+} + \uparrow^{8+} \\
\uparrow^{2+} \gg \uparrow^{8+} & \uparrow^{4+} + \uparrow^{5+} \gg \uparrow^{8+}.
\end{array}$$

(Here $A \gg B$ means that A exceeds any multiple of B), and these inequalities, together with the fact that $\uparrow^{n+}$ is always positive, suffice to imply all inequalities between sums of $\uparrow^{1+}, \ldots, \uparrow^{8+}$.

Using the Theorem of 17 October, this analyses every position of Bynum's game in which no side of any rectangle exceeds 16 (and many other positions).

The abbreviated list $\uparrow^{2+} \gg \uparrow^{3+} \gg \uparrow^{5+}$, $\uparrow^{2+} \gg \uparrow^{4+}$ suffices when no rectangle has a side greater than 10.

Our remark that a player can win if he has a lead of two heaps, or one heap and the move, means that in actual play one need only expand the situation into a sum of terms $\uparrow^{n+}$ when it is quite closely balanced. Then we use the assertion that a sum of terms $\uparrow^{n+}$ and $-\uparrow^{n+}$ will be positive if and only if it can be expressed as a sum of terms chosen from

$$\uparrow^{1+}, \uparrow^{2+}, \ldots, \uparrow^{2+} - N.\uparrow^{3+}, \uparrow^{2+} - N.\uparrow^{4+}, \ldots, \uparrow^{4+} + \uparrow^{5+} - N.\uparrow^{8+}$$

(corresponding to our list of basic inequalities), where the numbers N, which need not be the same, can be arbitrarily large.

So for instance in the sum

$$\uparrow^{2+} - \uparrow^{3+} + \uparrow^{4+} + \uparrow^{5+} - \uparrow^{6+} - \uparrow^{7+} - \uparrow^{8+}$$

we need a term $\uparrow^{2+} - N.\uparrow^{3+}$ to cope with the term $-\uparrow^{3+}$. Subtracting this, we have (for a different N)

$$N.\uparrow^{3+} + \uparrow^{4+} + \uparrow^{5+} - \uparrow^{6+} - \uparrow^{7+} - \uparrow^{8+}.$$

But now we need a term $\uparrow^{3+} + \uparrow^{4+} - N.\uparrow^{6+} - N.\uparrow^{7+}$ to account for the $-\uparrow^{6+}$, since we have no longer a term $\uparrow^{2+}$, and this leaves

$$N.\uparrow^{3+} + \uparrow^{5+} + N.\uparrow^{6+} + N.\uparrow^{7+} - \uparrow^{8+},$$

which is *not* positive, since no one of the basic inequalities can be used to eliminate the term $-\uparrow^{8+}$.

Of course such a sum can only be positive if the term $\uparrow^{n+}$ with the least n appears positively, and so the above sum is not negative either. It follows that in a position with this sum as value, the first player has the winning move. Such a position in Bynumbers is that where Left has heaps of sizes 2, 2, 5, 5, and Right heaps of sizes 1, 3, 3, 8. The reader might like to find winning moves for each of Left and Right as first player.

CHAPTER 16

The Long and the Short and the Small

. . . and there were present the Picninnies, and the Jobilillies, and the Garyulies, and the grand Panjandrum himself, with the little round button at top, and they all fell to playing the game of catch as catch can, till the gun powder ran out at the heels of their boots.
–Samuel Foote (printed in Maria Edgeworth's "Harry and Lucy Concluded")

This chapter discusses the ways in which long games (those with an infinity of positions) may differ from short ones. We start with a theorem which we have postponed from Chapter 9 so that it could serve as the text for a sermon.

THEOREM 90. *For every short game G which is not a number, we have the translation property*

$$G + x = \{G^L + x \mid G^R + x\}.$$

Proof. In the difference

$$\{G^L + x \mid G^R + x\} - x + \{-G^R \mid -G^L\}$$

we see that the moves in the components other than $-x$ have exact counters, and so we need only show that there is no good move in $-x$. But the move for Right to $-x^L$ takes us to the difference

$$\{G^L + x \mid G^R + x\} - (G + x^L) = H - K, \text{ say.}$$

But then H and K have Left and Right values obtained respectively by adding x and x^L to the Left and Right values of G, and so we cannot have $H \leqslant K$.

Now this theorem ceases to hold for all long games. Let $\mathbb{R}$ denote the set of all real numbers, and consider the games

$$A = \mathbb{R} \mid \mathbb{R}, \qquad B = \mathbb{R} \mid 0, \qquad C = \mathbb{R} \| \mathbb{R} \mid 0$$

We shall find that A, B, C have quite interesting properties. Of course it would make no difference if we were to replace $\mathbb{R}$ by $\mathbb{Z}$ (say) in their definitions.

The game $A = \pm\mathbb{R}$ plainly cannot have the translation property, since the set $\mathbb{R} + x$ is identically the same set as $\mathbb{R}$ for all real x, and so we should obtain the absurd equality $A + x = A$ for all real x. The trouble is that the Left and Right *values* of A do not exist, for the Left and Right *sections* are not near numbers.

We can still compute these sections using Theorem 56, with max and min replaced by sup and inf. We find $L(A) = \sup R(x)$, over all real x, so that $L(A)$ is the section, which it is natural to call ∞, between all real numbers and all larger numbers. In a similar way, we find $R(A) = -\infty$, $L(B) = \infty$, $R(B) = L(0)$, $L(C) = R(C) = \infty$.

So A is confused with all real numbers (and therefore with numbers between them), but lies strictly between any infinite negative number and any infinite positive one (say $-\omega^{1/\omega}$ and $\omega^{1/\omega}$). In a similar way, B is confused with real numbers greater than or equal to 0, but greater than all negative numbers and less than infinite positive ones. The game C is a little more interesting, since it is confused with no numbers, having all real numbers to its left and all infinite positive numbers to its right.

For G to have the translation property, it suffices that one of the two sections $L(G)$, $R(G)$ should be "next door to" some number y, that is to say, should have the form $L(y)$ or $R(y)$. So for instance, B has the translation property.

The temperature theory of Chapter 9 works wholesale for the Class of all games that have only finitely many stopping positions. We need only replace the words "dyadic rational" by "number" whenever they occur. For these games we can still draw thermographs, although the coordinate axes must have space for arbitrary numbers. In particular the mean value G_∞ always exists for such games, although for games with infinite temperature it can be a pretty useless concept.

Thus the game $\{1 + \omega \mid 1 - \omega\}$ has mean value 1, but since it is infinitely hot ($t = \omega$) it is not true that $m.G$ is necessarily very nearly $m.1$, and indeed for all odd m, $m.G = \{m + \omega \mid m - \omega\}$ is confused with all real numbers. With a slightly more complicated G we could make this hold for even m as well.

There are many more games for which the temperature theory works, but it is plainly not true that (for instance) every game has a mean value, in any reasonable sense. For since the game C lies between all real numbers and all positive infinite numbers, we should expect the same to be true of its mean value, which therefore could not be any number.

The game C is constructed in the same sort of way as the game considered in Chapter 10. In general, for any set S of numbers consider the game $G = S \| S \mid -x$, where $-x$ is less than every member of S. Then it is easy to see that the Left and Right sections of G are both the section $R(S)$ "just to the right" of S—that is to say, lying to the right of every number in S, but left of all greater numbers. Moreover, using Theorem 55 it is very easy to see that if H is any game greater than every number in S, then $G < H$ for all sufficiently large ordinals x. In particular, the positive game $+_{\alpha} = 0 \| 0 | -\alpha$ can be made less than any given positive game by choice of α, and so these games really do "tend to 0" as α "tends to **On**". In a similar way, the game $\mathbb{R} \| \mathbb{R} | -\alpha$ tends downward towards the real numbers as α tends to **On**.

We can use the same idea to investigate the largest infinitesimal games. Letting $\mathbb{R}^+$ denote the set of all strictly positive real numbers, we see that the game $1 | \mathbb{R}^+ \| \mathbb{R}^+$ is an infinitesimal game strictly greater than all infinitesimal numbers, and that if we were to replace 1 here by a large ordinal α we could obtain an "arbitrarily large" game with these properties.

GAMES OF ODD ORDER

It is easy to see that there can be no short game G satisfying the equation $G + G + G = 1$, for this equation implies that the mean value of G is $\frac{1}{3}$, whereas the mean value of any short game is a dyadic rational. On the other hand, there is a long G with this property, namely the number $G = \frac{1}{3}$. We shall see that there are others in a moment.

Now Theorem 89 of Chapter 15 shows us that there exist short games whose order is any desired power of 2, and Norton has generalised the construction we gave immediately after that theorem for producing submultiples of $\uparrow$, and deduced in particular that there are long games of all finite orders. We repeat his argument.

Let G be positive, and let H satisfy $H - K \geqslant$ all $G - G^R$, $G^L - G$, and $2K = G$, where $K = \{H | G - H\} > 0$. Then for numbers x define "multiples" $x.G$ (depending also on H) as follows. For x a finite integer, define $x.G$ to be the sum of the appropriate number of copies of G or $-G$. For other numbers $x = \{x^L | x^R\}$, define $x.G = \{x^L.G + H | x^R.G - H\}$. (When we want to emphasize the dependence on H, we can write $x_{(H)}G$. Thus for example we have $\hat{\frac{1}{2}} = \frac{1}{2}_{(\Uparrow *)}\uparrow$.) Then it is easy to see that $(x + y).G = x.G + y.G$, and hence in particular that $\frac{1}{3}.G = X$ satisfies the equation $X + X + X = G$. For $G = 1$, $H = \frac{1}{2}$, we find that $\frac{1}{3}_{(H)}1$ satisfies $X + X + X = 1$, but $X \neq \frac{1}{3}$, (since its Left and Right values are different), and so $X - \frac{1}{3}$ has order 3. Obviously we can construct games of any finite order like this.

But Norton has also proved what was quite a long-standing conjecture, that no *short* game has odd order. We follow his proof now. We call $G^L - G$ the *incentive* of the move from G to G^L, and $G - G^R$ the incentive of that from G to G^R, since these quantities measure the value of these moves to the player making them. Now we shall call G *balanced* if in whatever form G is taken, for each irreversible move from G there is a move for the other player of at least as great an incentive. (In fact it can be shown that G is balanced provided only that this condition holds for the simplest form, but we do not need this.)

THEOREM 91. *If G has finite order, then G is balanced.*

Proof. Suppose $n.G = 0$, and consider a typical option of G, say G^L. Since $G^L + (n-1).G \lhd\!\!\mid\!\mid 0$, we have either

$$\text{some } G^{LR} + (n-1).G \leqslant 0$$

or

$$\text{some } G^L + G^R + (n-2).G \leqslant 0.$$

In the first case the move to G^L is reversible, and in the second case the move to G^R has at least as great an incentive. The argument applies to every form of G.

It follows of course that if G is of finite order, then every multiple of G is balanced. The next theorem shows that any short game with the latter property has order some power of 2, and so, together with the previous theorem, establishes Norton's result.

THEOREM 92. *If G is born on day g, and $2^g.G$ is balanced, then $2^g.G = 0$.*

Proof. We suppose that all positions of G are in simplest form, and consider expressions of $2^g.G$ in the form

$$2^g.G = a.2^h.H + b.2^j.J + c.2^k.K + \ldots,$$

where $H, J, K, \ldots$ are distinct positions of G or $-G$ born on the respective days $h, j, k, \ldots$, and $a, b, c, \ldots$ are positive integers. We shall show that if any one of $H, J, K, \ldots$ is non-zero, the expression can be replaced by a simpler one of the same type, so that by repetition we may reduce the particular expression $2^g.G$ to 0.

We compare the simplicity of two such expressions E and F as follows. Enumerate the positions of G and $-G$ in any way which ensures that each position precedes all positions born on later days, and that any position H is adjacent to its negative $-H$, unless these coincide. Then call E simpler than F if the latest game of this list that occurs in just one of E and F is in F.

Now consider such an expression, with $H, J, K, \ldots$ not all zero. Then one of these games has a move (say the move from H to H^L) whose incentive is not strictly exceeded by the incentive of any other move from any of $H, J, K, \ldots$. If several moves have equal incentive, we suppose further that H is the possibility that appears latest in our list. Then the move from the given form of G to

$$H^L + (a.2^h - 1).H + b.2^j.J + \ldots$$

is either reversible or there is some move for Right of at least as great an incentive. Considering the various possibilities, we obtain an inequality of one of the forms:

$$H^L + J^R + (a.2^h - 1).H + (b.2^j - 1).J + \ldots \leqslant 2^g.G$$

$$H^L + H^R + (a.2^h - 2).H + b.2^j.J + \ldots \leqslant 2^g.G$$

$$H^{LR} + (a.2^h - 1).H + b.2^j.J + \ldots \leqslant 2^g.G$$

and so one of $H^L - H \leqslant J - J^R$, $H^L - H \leqslant H - H^R$, $H^{LR} \leqslant H$.

Since the move from H to H^L had maximal incentive, we must have equality in the first two cases, and since H is in simplest form, the third case cannot happen. But now we can simplify the given expression by replacing the term $a.2^h.H$ by

$$a.2^h.J^R + a.2^h.(-J) + a.2^h.H^L \quad \text{or} \quad a.2^{h-1}.H^L + a.2^{h-1}.H^R$$

in the two cases. The new expression is strictly simpler than the original, and still enjoys the property that the coefficient of any game born on day n is divisible by 2^n. (The strictness of the simplification follows from the fact that the games J^R, $-J$, H^L, H^R all appear strictly before H in our list.) As we remarked, we can now repeat the process to show that if $2^g.G$ is balanced, and in particular if G has finite order, then $2^g.G = 0$, showing that the order is a power of 2. We already know short games whose orders are arbitrarily chosen powers of 2, and some games of order 4 (of the form $x \mid * - x$ for numbers x) actually arise as positions in our games of dominoes and SNORT. It is probable that in these two games we can also find positions of orders 8, 16, . . . , etc.

Norton and I have slightly extended the argument of Theorem 92 so as to show that for any short game G and odd number n, G is expressible as an integral linear combination of the positions of $n.G$, in no matter what form $n.G$ is taken. So for instance there is no short game G with $3.G = \uparrow$, since no game of the form $a.\uparrow + b.*$ satisfies this equation.

The Class **No** of surreal numbers is defined as a Subclass of the Class **Pg** of games by the hereditary requirement that every game G in it satisfy $G^L < G < G^R$, for all G^L and G^R. Norton has also established a conjecture of mine that *any* game G for which $G^L < G < G^R$ holds for all G^L, G^R is already a number. In other words, any game other than a number is confused with some one of its options.

Perhaps the most significant way in which long games may differ from short is the lack of any theory of canonical forms for general long games. We showed in Chapter 10 that every short game had a unique simplest form, distinguished by having neither dominated nor reversible moves. Now for long games we certainly cannot hope to omit dominated moves, for example in the game $\omega = \{0, 1, 2, 3, \ldots \mid \}$ every move is dominated, and we certainly cannot omit them all! But of course we can omit certain infinite sets of moves without affecting the value—for instance we have

$$\omega = \{1, 2, 4, 8, 16, 32, \ldots \mid \}.$$

So we do not expect a simplest (i.e. smallest) form.

On the other hand, we can ensure that a quite arbitrary game has no reversible moves. We recall from Chapter 10 that the move from G to G^L is called reversible if we have some $G^{LR} \leqslant G$, and that we may then replace G^L by the set of all G^{LRL} (for this G^{LR}) without affecting the value of G. Now we may do this simultaneously for all reversible G^L (and G^R) and repeat until no move is reversible, and we are not led into an infinite regress, for this would entail an infinite sequence

$$G^L, G^{LRL}, G^{LRLRL}, G^{LRLRLRL}, \ldots,$$

and so an infinite play of G. The argument by which we proved Theorem 69 now proves that if G and H are free of reversible options, then we have $G = H$ iff each $G^L \leqslant$ some H^L, each $H^L \leqslant$ some G^L, each $G^R \geqslant$ some H^R, and each $H^R \geqslant$ some G^R, or in other words, two games without reversible options are equal if and only if each option of either is dominated by a corresponding cption of the other.

It does not seem to be possible to do better. In particular, Norton has disproved a fairly long-standing conjecture (the *ancestors* conjecture) by producing two forms for a certain long game G with the property that G cannot be expressed in terms only of the positions common to both forms. This cannot happen for a short game G, since the theory of Chapter 10 shows that then every position in the simplest form of G arises as the value of a position in every form of G. Norton's game is

$$G = \{0 \,\|\, 0 \mid 0, 0 \mid -2, \ldots\} \quad \text{and} \quad \{0 \,\|\, 0 \mid -1, 0 \mid -3, \ldots\}.$$

The following result was promised in Chapter 15. Although it refers to all games its main applications seem to be in comparing the sizes of very large or very small games, so we give it here. In particular, it has an important application to the calculus of atomic weights which we shall describe shortly. We say that "G involves X" to mean that some position of G has value X. Recall that $X : Y$ denotes the ordinal sum, defined inductively by:

$$X : Y = \{X^L, X : Y^L \mid X^R, X : Y^R\}.$$

THEOREM 93 (Norton's lemma). *X and $X : Y$ have the same order-relations with all games G not involving X.*

Proof. Since for any Y there is some ordinal α with $-\alpha \leqslant Y \leqslant \alpha$ (Theorem 55), we are reduced to proving that X and $X:\alpha$ (or $X:-\alpha$) have the same order-relations with all games G not involving X. Taking obvious inequalities into account, this amounts to proving that

$$X:\alpha \geqslant G \text{ implies } X \geqslant G \quad \text{and} \quad X \leqslant G \text{ implies } X:\alpha \leqslant G$$

for every positive ordinal α. From the definition, we have

$$X:\alpha = \{X^L, X:\beta \mid X^R\},$$

where β ranges over the ordinals less than α.

Suppose first that $X:\alpha \geqslant G$, but $X \lhd\!| \; G$. Then there must be a good move for Right from $X - G$. But his move to $X^R - G$ is also available from $X:\alpha - G$, and so this must be to $X - G^L$, say. Since X is not involved in G, it cannot be involved in G^L, and so from $X \leqslant G^L$ we can deduce inductively that $X:\alpha \leqslant G^L$, contradicting $X:\alpha \geqslant G$.

So we must suppose that $X \leqslant G$, but $X:\alpha \; |\!\rhd G$. What is Left's good move from $X:\alpha - G$? Certainly not to $X^L - G$, for this is available also from $X - G$; and not to $X:\alpha - G^R$, for this implies $X:\alpha \geqslant G^R$, so inductively $X \geqslant G^R$, contradicting $X \leqslant G$. So we must suppose that Left's good move is to $X:\beta - G$ for some ordinal $\beta < \alpha$. But now we have $X:\beta \geqslant G$, and so inductively $X \geqslant G$, which combines with the assumption $X \leqslant G$ to show that $X = G$ is involved in G.

Before we proceed to the applications to small games we deduce the corollary promised in Chapter 15.

THEOREM 94. *If neither X nor Y has a reversible move, then neither does $X:Y$.*

Proof. Suppose to the contrary that $X:Y$ has a reversible move, say for Left. What is this move?

If it is a move to X^L, reversed to X^{LR}, we have $X^{LR} \leqslant X:Y$, so by Norton's lemma $X^{LR} \leqslant X$, since X^{LR} does not involve X, showing that after all X had a reversible move to X^L.

If it is to $X:Y^L$, reversed to X^R, we have $X^R \leqslant X:Y$, which is impossible since X^R is a Right option of $X:Y$.

Finally, if it is to $X:Y^L$, reversed to $X:Y^{LR}$, we have $X:Y^{LR} \leqslant X:Y$. But from Chapter 15 we know that $X:Y$ is a sop function (strictly order preserving) in Y, so we must have $Y^{LR} \leqslant Y$, showing that after all Y had a reversible move to Y^L.

Now we know that in general the value of $X:Y$ depends on the form of X, rather than only on its value. But we can use our earlier remarks to select an 'absolute' version, defined as $X:Y$ for forms X and Y without reversible options. It is easy to see that if X_1 and X_2 are forms of the same game neither of which has a reversible option, then $X_1 : Y = X_2 : Y$, for each option of any of these is dominated by a corresponding option of the other. Theorem 94 now further shows that $X:(Y:Z) = (X:Y):Z$ under this convention, for no new reversibility is introduced on either side. Of course, since $X:Y$ is a sop function of Y, its value depends only on the value, and not the form, of the second variable Y.

THE GAMUT OF GAMES

We are now sufficiently well-informed to present a fairly complete picture of the possible types of magnitude for positive games.

To fix our ideas, we first discuss the possible sizes of numbers.

very small: numbers like $\frac{1}{\omega}, \frac{1}{\omega^2}, \ldots, \frac{1}{\alpha}$ for large ordinals α;

fairly small: numbers like $\frac{2}{\omega}, \frac{3}{\omega}, \frac{1}{\sqrt{\omega}}, \frac{1}{\sqrt[3]{\omega}}, \ldots, \frac{1}{\omega}, \omega^{-\alpha}$;

ordinary sized: numbers like $\frac{1}{100}, \frac{1}{2}$, 1, 2, 100;

fairly large: numbers like $\frac{1}{2}\omega, \sqrt{\omega}, \omega^{1/\omega}, \omega^{\omega^{-\alpha}}$; *and*

very large: numbers like $\omega, \omega^2, \ldots, \omega^\omega, \alpha$ for large ordinals α.

If α is a large ordinal we can say that the smallest infinitesimal numbers are like $1/\alpha$, and the largest like $1/(\omega^{1/\alpha})$, while the smallest infinite numbers are like $\omega^{1/\alpha}$ and the largest like α.

When we add general games the above scale needs enlarging at several points. So we now consider:

GAMES IN THE GAPS

In Chapter 3 we discussed the gaps in the number line. Only some of these gaps can contain games, since it follows from the discussion in Chapter 9 that the gaps $L(G)$ and $R(G)$ must each be upper or lower bounds of non-empty *sets* of numbers. Every gap with this property has the form $x + \omega^\Xi$ or $x - \omega^\Xi$, where x is a number and Ξ another gap that is the upper or lower bound of a set. But now the gap Ξ may be **On** or $-$**On**, which do not contain games, since they are the lower and upper bound of the *empty* set.

These remarks show that it suffices to discuss the games that lie in gaps of the form ω^Ξ, which have the property that the sum of any two games from the gap is again in the gap. The particular case $\Xi = -$**On**, when ω^Ξ is the gap $1/$**On** containing small games, is rather special and will be discussed later. Otherwise Ξ is either the upper or lower bound of a non-empty set, which gives two cases in the argument.

The gap ∞ typifies the first case. The smallest games in this gap have already been described—they are the games $\infty_\alpha = \mathbb{R} \| \mathbb{R} | -\alpha$ for large ordinals α. The largest games present rather more of a problem. It turns out that they are the games ∞^α defined inductively by

$$\infty^1 = \infty = \mathbb{R} \| \mathbb{R} | \mathbb{R}$$
$$\infty^\alpha = \{\infty^\beta . n \| \infty^\beta . n | -\infty^\beta . n\} \quad \text{if } \alpha \geqslant 2$$

where β ranges over all the ordinals less than α, and n over all the positive integers.

There are some interesting identities between these games. If x is a finite number, then $\infty_{\alpha+x} = \infty_\alpha - x$. But the most interesting results concern the game ∞ itself. We have $\infty + \infty = \infty_0$ (the game C of our introduction to this chapter), and also the equalities

$$\mathbb{R} | \mathbb{R} = \pm\infty, \quad \mathbb{R} | 0 = \infty | 0 = (2.\infty) | 0 = \infty \pm \infty,$$
$$\infty | \pm\infty = \infty +_\infty, \quad \infty \| \infty | 0 = 2.\infty.$$

Some similar equalities exist involving higher powers of ∞^α.

Now let Ω be another gap of the form ω^Ξ ,where Ξ is the upper bound of a non-empty set. Then we can define games

$$\Omega_\alpha = \{\omega^x . n \| \omega^x . n | -\alpha\}_{x \in S}$$
$$\Omega^1 = \{\omega^x . n \| \omega^x . n | -\omega^x . n\}_{x \in S,\, n=1,2,3,\ldots}$$
$$\Omega^\alpha = \{\Omega^\beta . n \| \Omega^\beta . n | -\Omega^\beta . n\}_{\beta<\alpha,\, n=1,2,3,\ldots},$$

and it turns out that the Ω_α are the smallest games in Ω, and Ω^α the largest.

The gap $1/\infty$ typifies the other major case. We can define

$$(1/\infty)_\alpha = \{\alpha | \mathbb{R}^+ \| \mathbb{R}^+\}$$

where $\mathbb{R}^+$ denotes the set of positive reals, and these games are the largest in $1/\infty$. To find the smallest we need a more complicated construction, defining games

$$\infty^{-1}(x) = \{x | \mathbb{R}^+ \| \mathbb{R}^+ \||| \mathbb{R}^+\}$$
$$\infty^{-\alpha}(x) = \{x | \infty^{-\beta}(y) \| \infty^{-\beta}(y) \||| \infty^{-\beta}(y)\}_{y \in \mathbb{R}^+,\, \beta<\alpha}$$

where α is any positive ordinal and x any number. Then it turns out that the games $\infty^{-\alpha}(x)$, or even just $\infty^{-\alpha} = \infty^{-\alpha}(1)$, are the smallest games in $1/\infty$.

In the general case in which $\Omega = \omega^\Xi$, and Ξ is the lower bound of a non-empty set S, we can define

$$\Omega_\alpha = \left\{\alpha \,\middle|\, \omega^s.\frac{1}{n} \,\middle\|\, \omega^s.\frac{1}{n}\right\}_{s\in S}$$

$$\Omega_1(x) = \left\{x \,\middle|\, \omega^s.\frac{1}{n} \,\middle\|\, \omega^s.\frac{1}{n} \,\middle|\middle|\middle|\, \omega^s.\frac{1}{n}\right\}_{s\in S}$$

$$\Omega_\alpha(x) = \{x \mid \Omega_\beta(y) \,\|\, \Omega_\beta(y) \,|||\, \Omega_\beta(y)\}_{y>\Omega,\, \beta<\alpha}.$$

(Since Ω is the lower bound of a set, the impropriety in letting y range over all numbers greater than Ω is only apparent). I do not know why $\Omega_\beta(y)$ must appear three times in this definition.

The case $+ = 1/$**On** is special. However, it is easily solved. We have already shown that $+_\alpha = 0 \| 0 \mid -\alpha$ are the smallest games in this gap. The largest are the games $\infty^\alpha.\uparrow$, the multiples of $\uparrow$ by the largest games in the gap ∞. This remark shows perhaps most conclusively that there is no natural definition of $x.\uparrow$ if x is an infinite *number*, for there is no game greater than all the games $\infty^\alpha.\uparrow$ and less than all positive numbers.

THE GAMUT REVEALED

We use x for a positive number, possibly further restricted, and α for an arbitrary ordinal. Then we have:

The very smallest games $+_1, +_2, \ldots +_\infty, \ldots +_\alpha$.
The smallest all smalls $+_{x.\uparrow}$ (x finite). $+_{(\infty^\alpha.\uparrow)}$
The $\uparrow$ *scale* $\uparrow, \uparrow^2, \uparrow^3, \ldots, \uparrow^\omega, \ldots, \uparrow^\alpha$.
Largest below $\Uparrow$ $*01, *012, *0123, \ldots *012\ldots\beta\ldots (\beta < \alpha)$ (defined soon).
Multiples of $\uparrow$ $\uparrow, \Uparrow, \ldots, x.\uparrow$ (finite x).
Largest small games $\infty.\uparrow, \infty^2.\uparrow, \ldots, \infty^\alpha.\uparrow$.
The smallest infinitesimal numbers $1/\omega, \ldots, 1/\alpha$.
The largest infinitesimal numbers $2/\omega, 1/\sqrt{\omega}, \ldots, 1/\omega^{1/\alpha}$.
The next smallest games $\infty^{-1}, \infty^{-2}, \ldots, \infty^{-\alpha} = \infty^{-\alpha}(1)$.
The largest infinitesimal games $\alpha \mid \mathbb{R}^+ \| \mathbb{R}^+ = (1/\infty)_\alpha$.
The finite numbers $\frac{1}{100}, \frac{1}{2}, 1, 2, 100, \ldots$.
The smallest infinite games $\infty_\alpha = \mathbb{R} \| \mathbb{R} \mid -\alpha$.
The largest games in ∞ $\infty^\alpha = \infty^\beta.n \| \infty^\beta.n \mid -\infty^\beta.n$.
The smallest infinite numbers $\omega/2, \sqrt{\omega}, \omega^{1/\omega}, \ldots, \omega^{1/\alpha}$.
The largest infinite numbers $\omega, \omega.2, \ldots, \omega^\omega, \ldots, \omega^\alpha$.

We note that it is the Archimedean principle that tells us that the numbers α really do "tend to **On**" in the sense that every game is less than all sufficiently large ordinals. It is a fairly easy deduction that $+_\alpha$ really does "tend to 0" in the sense that any positive game is greater than some $+_\alpha$.

It is also easy to see that if x and y are numbers with $x < y$, then for the corresponding tiny games we have $+_y$ infinitesimal with respect to $+_x$, so that in these games we have a scale of infinitesimals. And we can also let x vary over the infinite numbers less than ω, or over some class of games between the finite and infinite numbers, so that even the tiny world has a very rich structure. The smallest all small games have the form $+_{(x\,.\,\uparrow)}$ where x is a finite number if we want a short game, but may be ∞^α if we allow long ones. While we are still on the subject of tiny games, we might remark that it is amusing to verify that for *any* game G, we have $+_{+_{+_G}} = \uparrow$, so that in particular, $\uparrow$ is the unique solution of $G = +_G$.

We can use Norton's lemma to show that the game $\dot{+}_x$ is smaller than any positive all small game G, as follows. We find that $\{x\,|\,0\} : 1$ can be expressed as $\{x|0\} + {}_x$, so that $\{x|0\}$ and $\{x|0\} + {}_x$ have the same order-relations with $\{x|0\} + G$, since this does not involve $\{x|0\}$.

THE SUPERSTARS, AND THE GAME OF SUPERNIM

The games $*abc\ldots$, which we call the *superstars*, are defined by the equation

$$*abc\ldots = \uparrow * + \{*a, *b, *c, \ldots \,|\, *0, *1, *2, \ldots\},$$

where the sequence $0, 1, 2, \ldots$ on the right is long enough to include the mex of all the numbers $a, b, c, \ldots$ on the left. If S is the set of numbers $a, b, c, \ldots$ we sometimes write $*S$ for $*abc\ldots$. These games arise naturally in several places, and have the following properties

(i) If S has a single element s, say, then $*S = *s$ as usually defined.

(ii) If S is a proper subset of T, then $*S < *T$.

(iii) We have $*n \leqslant *S$ iff $n \in S$ (and otherwise $*n \,\|\, *S$).

(iv) We have $*S \leqslant \uparrow * + *n$ iff $n \notin S$ (and otherwise $*S \,\|\, \uparrow * + *n$).

There is also a *restricted translation-invariance* property—if the set T has the form $S +_2 n$ for *some* n, then $*T = *S + *n$ for the least such n. (Here $S +_2 n$ denotes $\{s +_2 n \,|\, s \in S\}$.)

A *supernim* position is a game of the form

$$G = \{*a, *b, *c, \ldots \,|\, *\alpha, *\beta, *\gamma, \ldots\}$$

which has both Left and Right options which are all Nim-heaps. For such games we have the identities

$G = *m$ if both sets $S = \{a, b, c, \ldots\}$ and $\Sigma = \{\alpha, \beta, \gamma, \ldots\}$ have the same mex m.

$G = \downarrow * + *S$ if S has the smaller mex.

$G = \uparrow * + *\Sigma$ if Σ has the smaller mex.

So every supernim position is either equivalent to a Nim-heap or to one of form $\{*a, *b, *c, \ldots \mid *0, *1, *2, \ldots\}$ or $\{*0, *1, *2, \ldots \mid *a, *b, *c, \ldots\}$ in which we may suppose that one side or the other contains all numbers we like. We call the former type Right's terms and the latter Left's ones.

We assert that in a sum of such terms (including Nim-heaps) Left will win if he has at least two more terms than Right, or at least one more and the advantage of having first move. He wins in fact by destroying Right's terms as quickly as possible (replacing them by arbitrary Nim-heaps). When it is his move and he has destroyed them all, there will still be at least one Left term left. If there is exactly one such, Left can replace it by a Nim-heap of just such a size as to make the resulting Nim-position have value zero. If there are just two such terms, Left can replace one by a Nim-heap so large that Right (who is restricted in his choice for the other) cannot replace the other by one which makes the Nim-game have zero sum. Finally, if there are three or more Left terms remaining, Left can afford to replace any one of them by an arbitrarily chosen Nim-heap and leave further decisions to his next move.

This argument proves that any sum or difference of terms $*S$ is infinitesimal with respect to $\Uparrow$. We shall use this property later. The remaining cases, in which the mover has one less term than his opponent, present a more difficult problem. But if each player follows the policy of destroying his opponent's terms, his opponent can do no better than to do likewise, and the game reduces to the game of *supernim.*

In this game, each player has a number of cards, each card being labelled with a number of Nim-heaps it may be exchanged for. It does no harm to suppose that initially the two players have the same number of cards. The players then alternately *declare* their cards, replacing them by one of the permitted Nim-heaps. When all cards have been declared, they play the resulting Nim-game to decide the winner. (Notice that in this translation Left's terms have become Right's cards).

The sum $*S + *T + \ldots - *U - *V \ldots$ has then the same outcome as the supernim game in which Left has cards corresponding to the sets $S, T, \ldots$ and Right cards corresponding to $U, V, \ldots$. In particular we see that the outcome of such a sum is unchanged when we replace the numbers appearing in the sets $S, T, \ldots, U, V, \ldots$ by any other numbers with the same Nim-sum relations. There does not appear to be a complete theory, and we shall consider in detail only the case when the numbers appearing are chosen from $0, a, b, c$, where $c = a +_2 b$.

In this case the restricted invariance principle allows us to suppose every term is one of $*0a$, $*0b$, $*0c$, $*abc$, or $*0abc$ (or their negatives) together with Nim-heaps $*0$, $*a$, $*b$, $*c$ which we can combine if we like. (For most of the game they have a rather negligible effect.)

Then a term $*0abc$ *is greater than any sum of other terms.*

(For a player with a card bearing all labels 0, a, b, c obviously wins by declaring it to have the appropriate value right at the end of the game: we suppose of course by cancellation that his opponent has no such card.)

If there is no term $*0abc$, *the terms* $*abc + *0a$ *(say) beat any combination of other terms.*

(For a player holding the two corresponding cards can at his next to last declaration choose 0 or a so as to ensure that his last declaration wins the Nim-game.)

A term $*0a$ *is not beaten by any sum of terms* $*0b$, $*0c$ *and Nim-heaps.*

(For the player holding it can at his last declaration choose 0 or a so that his opponent (who must declare 0 or b say) cannot correct the resulting Nim-game to have zero value.)

Finally, the term $*0a$ *(or any multiple thereof) beats only the Nim-heaps* $*0$ *and* $*a$.

Since we can always add terms $*0$ so as to balance the number of terms on each side, the above theory handles all supernim games in which all labels are chosen from 0, a, b, c, and all sums or differences of such superstars.

The superstars play an important role in the atomic weight calculus, to which we now proceed, since they include the largest games of atomic weight zero.

THE THEORY OF THE SMALL WORLD. ATOMIC WEIGHTS

Perhaps this is the most useful and intriguing topic of this chapter; the theory of magnitude in the small world, developed jointly by Norton and the author. We use the term *small world* for the large family of games whose sizes are most naturally measured in units of $\uparrow$. The small world behaves in many respects like the large one, but often with a fundamental "uncertainty" of size about $\uparrow$ or $\Uparrow$ which makes exact calculation rather difficult. The main achievement is the calculus of atomic weights, measuring small games in terms of $\uparrow$, which enables us to use the mean-value calculus in the small world.

Because the complete theory is rather difficult and does not yet seem to be in final form, we omit several proofs. Norton hopes to present a complete account of this theory and his extensions of it in due course.

COMPARING G WITH THE STARS. REMOTE STARS AND ATOMIC WEIGHTS

We need to be able to decide for which n we have $*n \geqslant G$ or $*n \leqslant G$. Accordingly we define the *Above* and *Below* classes of G by:

$$A(G) = \text{class of } n \text{ with } *n \geqslant G$$
$$B(G) = \text{class of } n \text{ with } *n \leqslant G.$$

THEOREM 95. *Suppose the Above and Below classes known for all options of* G. *Define* A *as the class of* n *not in any* $B(G^L)$, *and* B *as the class of* n *not in any* $A(G^R)$. *Then if* A *has the smaller minimum* (*or* B *is empty*), *we have* $A(G) = A$ *and* $B(G)$ *empty, while if* B *has the smaller minimum or* A *is empty we have* $A(G)$ *empty and* $B(G) = B$. *Finally, if* A *and* B *have the same minimum* m, *then we have* $G = *m$, *and so* $A(G) = B(G) = \{m\}$.

(The proof is a fairly easy calculation.)

It follows from this algorithm that the comparison of G with $*n$ is ultimately independent of n. Either for all sufficiently large n we have $G > *n$ or for all sufficiently large n $G < *n$, or G is incomparable with all sufficiently large $*n$. For short games G the comparison becomes constant for n larger than the number of moves in G, but for long games we might need an infinite ordinal number for n. But in any case we can speak of comparing G with *the remote stars.* It turns out that the result of this comparison plays an important role in determining the atomic weight of G—this fact we might call *Mach's principle for the small world.* For the sake of precision we shall call *$*n$ a remote star* if $*n$ is not involved in G, and suppose G a short all small game.

Then the exact form of Mach's principle is that *the atomic weight of* G *is at least* 1 *if and only if* G *exceeds the remote stars.* Since we have not yet defined atomic weights, we interpret this for the moment in the form:

THEOREM 96. *If* G *exceeds the remote stars, then for any finite* N *the sum of* $N + 2$ *copies of* G *exceeds the sum of* N *copies of* $\uparrow$.

Proof. Let $*n$ be a remote star. Then if $G > *n$ we have $G > (*n){:}1$ by Norton's lemma. But

$$(*n){:}1 = \{*0, *1, \ldots, *n \mid *0, *1, \ldots, *(n-1)\} = \uparrow * - *01\ldots(n-1).$$

The desired assertion now follows from the analysis of supernim.

THE SUPERCOOLING FUNCTION G^t

Let t be a real number satisfying $1 < t \leqslant 2$. Then there is a supercooling function G^t defined for all short all small games G. This is defined in exactly the same way as the ordinary cooling function G_t of Chapter 9, but with a different and more subtle proviso. In fact G^t is defined by the inductive definition

$$G^t = \{G^{Lt} - t \mid G^{Rt} + t\}$$

except possibly when there is more than one *permitted integer* N, that is to say, more than one N which satisfies

$$G^{Lt} - t <\| N <\| G^{Rt} + t$$

for all G^L, G^R.

In this excepted case, G^t is defined as:

the largest nermitted integer (necessarily positive) if G exceeds the remote stars

the least permitted integer (necessarily negative) if G is exceeded by the remote stars

the integer zero (necessarily a permitted integer) if G is incomparable with the remote stars.

To summarise, we may say that G^t is defined by the inductive formula $G^t = \{G^{Lt} - t \mid G^{Rt} + t\}$ except that when we are faced with a choice between several integers we choose not the *simplest*, but rather the *greatest*, *least*, or the number *zero* according to the remote star criterion. The amazing result, which we do not prove here is:

THEOREM 97. *The supercooling function is a homomorphism* (*from short all small games to games*). *In other words, we have* $(G + H)^t = G^t + H^t$, *and* $(-G)^t = -G^t$.

There are some difficulties in defining a supercooling function for other values of t. For $t > 2$ there is no difficulty, however. We can in fact define G^t consistently for all $t > 1$ by use of the equation $G^{t+u} = (G^t)_u$, which will hold whenever both sides are defined. For $0 < t \leqslant 1$ there are several alternative definitions which achieve the same result, but no one is particularly satisfactory, and there are even worse problems for $t = 0$.

The most important case is $t = 2$, and so we write G'' for G^2, and call G'' the *atomic weight* of G. In fact the atomic weight determines G^t for all $t > 1$ (for $t \geqslant 2$ by cooling, and for $t < 2$ by 'heating up'), so that we do not need G^t for any t other than 2. The following omnibus theorem collects atomic weight information:

THEOREM 98. (i) *We have* $G'' = \{G^{L''} - 2 \mid G^{R''} + 2\}$ *except that when this* *permits more than one integer, G'' is the largest or least permitted integer or is zero according as $G > *n$, $G < *n$, $G \parallel *n$* for remote $*n$.

(ii) *Atomic weights are additive—so* $(G+H)'' = G'' + H''$, $(-G)'' = -G''$,

(iii) $G'' \geqslant 1$ iff *G exceeds remote stars.*

(iv) $G'' \leqslant -1$ iff *G is exceeded by remote stars.*

(v) *If* $G'' \geqslant 2$, *G is positive, and if* $G'' \leqslant -2$, *G is negative.*

(vi) *If* $G'' \,|\!\!\triangleright\, 0$, *then* $G \,|\!\!\triangleright\, 0$, *and if* $G'' \triangleleft\!\!| \, 0$, *then* $G \triangleleft\!\!|\, 0$.

(vii) *For* $G = \hat{x} = x.\uparrow$, *we have* $G'' = x$. *In particular, the atomic weight of* $\uparrow$ *is* 1. *Also the atomic weight of* $\uparrow x$ *is* 1 *for* $x > 0$.

(viii) *The atomic weights of* $\uparrow^x$ $(x > 1)$, $*n$, $*abc\ldots$, *are all zero.*

Let us compute the atomic weight of the game $\{0 \mid \uparrow, \uparrow *\}$ which we called $\Uparrow 1;1$ in Chapter 15. Here G^L has weight 0, and each G^R has weight 1, and so the inductive formula gives $\{0 - 2 \mid 1 + 2\}$, which permits the three integers 0, 1, or 2. So the atomic weight of G will be 0 or 2 according as G is incomparable with or exceeds the remote stars. So we must use our method for comparing G with the stars.

Here we have $B(G^L) = \{0\}$, and so A is the complementary class $\{1, 2, 3, \ldots\}$. Again, $A(G^R)$ is empty for each G^R, since we have no $*n \geqslant \uparrow$ or $\uparrow *$, and so the class B contains all integers, $B = \{0, 1, 2, \ldots\}$. Since B has the smaller minimum (0), it "wins", and we have $A(G)$ empty, $B(G)$ "full", so that in fact G exceeds all stars, and in particular, the remote ones. So the atomic weight of G is the largest permitted integer 2.

Now let us consider the game $\{\Uparrow \mid \downarrow *\}$. Here the inductive formula reads $\{2 - 2 \mid -1 + 2\} = \{0 \mid 1\}$, and since there is no integer between 0 and 1 this already gives the correct answer $\frac{1}{2}$. So atomic weights need not be integers. The example $\{\Uparrow \mid \Downarrow\} = \pm \Uparrow$ shows that they need not even be numbers, for here the inductive formula gives $\{2 - 2 \mid -2 + 2\} = \{0 \mid 0\} = *$ for the atomic weight. In fact we saw in Chapter 15 that in a natural sense the product of $\uparrow$ and $*$ is $\pm \Uparrow + *$, and so on taking atomic weights we have

$$* = (\pm \Uparrow)'' + (*)'',$$

whence indeed $(\pm \Uparrow)'' = *$.

Nevertheless the atomic weights of simple games tend to be numbers, and even integers, since after all the atomic weight is a kind of cooling function. This nice property is of course counterbalanced by a nasty one—we cannot assert that a game of positive atomic weight is positive, but only that it is positive or fuzzy. To achieve positivity of G we must know that G's atomic weight is at least 2, although if its weight is 1 or more we *can* assert that G exceeds almost all stars.

ATOMIC MASS THERMOGRAPHY

The atomic thermograph is a device for determining atomic weights, or to be more precise, *atomic mean weights* G^∞ (the mean value of G'', or equivalently the value of G^t for large t). Since for the inductive calculation we need to compare G with the stars, we shall draw the atomic thermograph like the ordinary thermograph, but write the members of $A(G)$ to its left, and those of $B(G)$ to its right (Fig. 62):

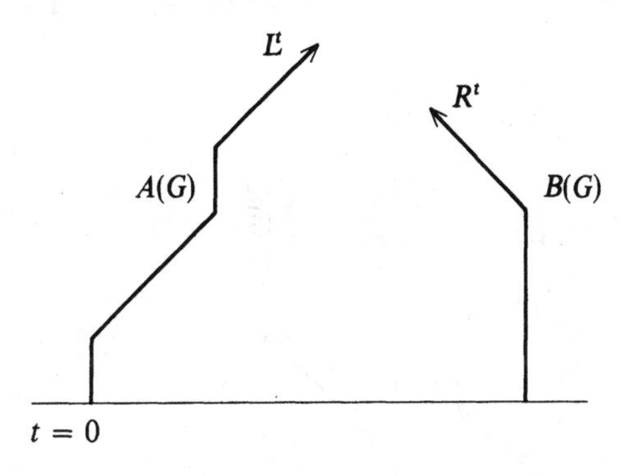

FIG. 62.

Given atomic thermographs for the options of G we compute that for G as follows. We adopt notation and conventions similar to those of Chapter 9, which the reader should consult in conjunction with this one. The atomic thermograph has Left and Right boundaries $Ł^t(G)$ and $R^t(G)$ which are computed by a rather more complicated rule than the boundaries $L_t(G)$ and $R_t(G)$ of the ordinary thermograph for a large game G.

Tentatively we define

$$Ł^t = \max R^t(G^L) - t, \quad R^t = \min Ł^t(G^R) + t.$$

We obtain in this way two curves which are our first approximation to the Left and Right boundaries. As t increases, the Left curve $Ł$ ultimately tends diagonally rightward (i.e. $Ł^t$ *decreases*), and the Right one diagonally leftward (R^t increases). We must examine these curves at the level $t = 2$. If the Left curve is in fact to the left of the Right one at this critical level (i.e. $Ł^t \geqslant R^t$ at $t = 2$), then they do indeed define the Left and Right boundaries of the atomic thermograph until they meet, from when both boundaries coincide in a single vertical line (the mast).

If however the Left curve is to the right of the Right one at the critical level (i.e. $Ł^t < R^t$), then the atomic weight is a number, and the atomic thermograph consists entirely of a mast at this number. In this case G'' is the simplest permitted number (i.e. number x satisfying $Ł^t < x < R^t$ at $t = 2$) unless there are at least two permitted *integers*, when G'' is that integer determined by the remote star criterion. Of course to compare G with the remote stars involves comparing it with all stars, and so computing the sets $A(G)$ and $B(G)$, which we enter on the diagram as we compute it. Each such set either contains almost all or almost no numbers, and it is this dichotomy which settles the atomic weight of G in the critical case.

As an example we illustrate (fig. 63) the calculation of the atomic thermograph for the game $G = \{3.\uparrow | \uparrow, \pm\Uparrow\}$. Here the thermographs for the options

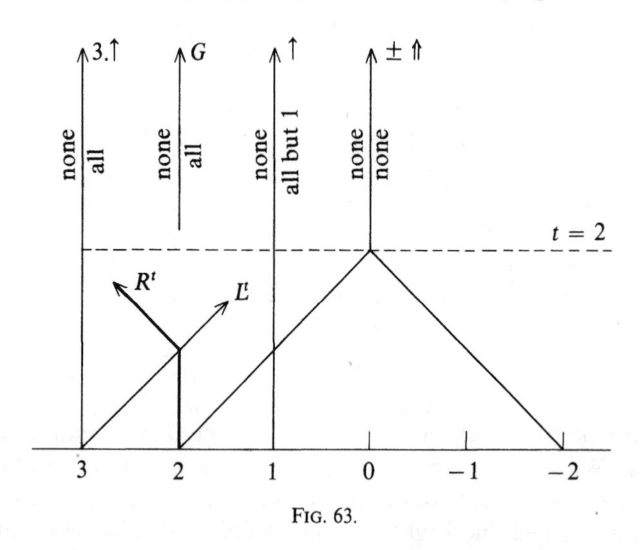

Fig. 63.

are supposed given, and the Left and Right curves L^t and R^t cross below the critical level $t = 2$. Since 2 is the simplest number between them at this level, and is the only integer between them, the atomic weight is 2 and the thermograph a mast at 2, with $A(G)$ empty, $B(G)$ full (this is indicated by the words "none", "all" to left and right of the mast).

In Figs 64 and 65 we show the atomic thermographs for the games which result when $3.\uparrow$ is replaced by $\uparrow$, $\uparrow *$, $4.\uparrow$, and $6.\uparrow$. In the first two cases there are three permitted integers 0, 1, 2. But in the first case we have $A = \{1\}$

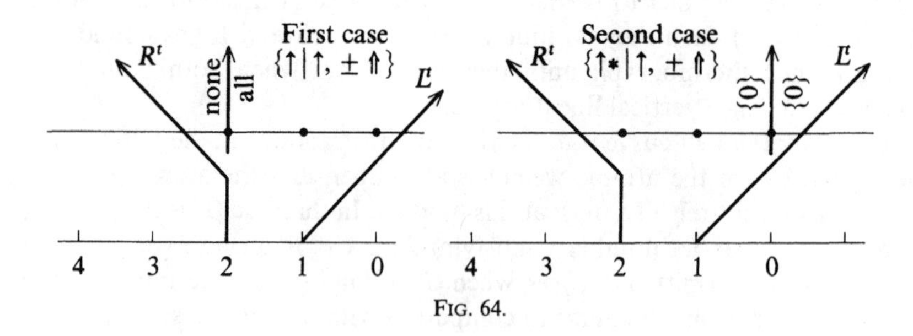

Fig. 64.

and in the second $A = \{0\}$, while $B = \{\text{all}\}$ in every case. So in the first case we have $A(G)$ empty, $B(G) = B$ full, while in the second case $A(G) = B(G) = \{0\}$ since then A and B have the same minimum. So in the first case the remote star criterion gives $G'' = 2$, and in the second case $G'' = 0$.

In the third case we have $L^t = 2$, $R^t = 3$ at $t = 2$, and so G'' is $2\frac{1}{2}$ by the simplest number criterion. In the fourth case $L^t = 4$, $R^t = 3$ at $t = 2$, and so the thermograph has a pyramid shape. We should point out that we have assigned no meaning to that part of the atomic thermograph corresponding to $t \leqslant 1$, but it seems that the calculation is easier if we draw this part nevertheless. [For $t > 1$ the atomic thermograph at level t indicates the Left and Right values of the supercooled game G^t.]

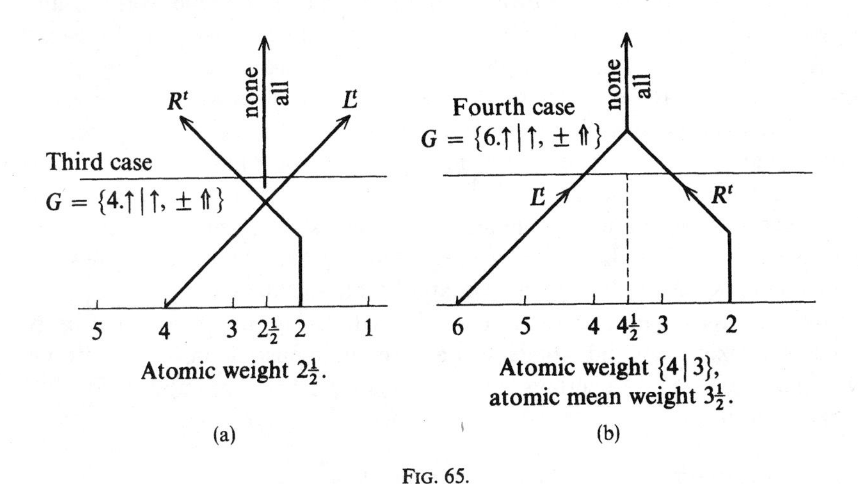

FIG. 65.

In practice, calculation with atomic weights tends to be easier than the above account might suggest, because the additional cooling effect tends to make atomic thermographs simpler than ordinary ones for games of roughly the same complexity.

Berlekamp and Wolfe have made effective use of another search operation, "chilling," in their theory of Go. They have also shown that this and various notions of "heating" and "overheating" arise naturally in Domineering and other games. See *Mathematical Go: Chilling Gets the Last Point*; "Blockbusting and Domineering" in the *Journal of Combin. Theory Ser. A;* and "Introduction to Blockbusting and Domineering," from *The Lighter Side of Mathematics.*

We conclude with a remark justifying our placing of the superstars in the gamut of games.

THEOREM 99. *Any short all small game G which has atomic weight zero is infinitesimal with respect to* $\Uparrow$ *and dominated by some superstar.*

Proof. Since multiples of G also have atomic weight zero, they are less than $\Uparrow$. Also $H = \Uparrow* - G$ has atomic weight 1 and therefore exceeds some remote star$*n$. It follows from Norton's lemma that $H > (*n)$: 1, and this is equivalent to the assertion $G < *01 \ldots (n-1)$.

The statement refers to $\Uparrow$ rather than to $\uparrow$, because the superstars themselves are infinitesimal with respect to $\Uparrow$ but not with respect to $\uparrow$, since in fact $*abc \ldots$ is incomparable with $\uparrow$ if 1 appears among the numbers $a, b, c, \ldots$.

With this we conclude our investigations into the remarkable world of ups, downs, stars, and superstars, observing that we have shown that this small world behaves very like the large one, but with some extra subtleties all its own. It seems that the theory of atomic weights is about as complete as we could expect for games which are naturally compared to $\uparrow$. But perhaps we can hope for an extended version of this theory which would enable us to measure still smaller games in terms of $\uparrow^2$, $\uparrow^3$, and so on. But the curious and complicated nature of the atomic weight algorithm suggests that any such theory will be very difficult to find. What do we expect to play the role of the remote stars, which enter so mysteriously and essentially into our theory?

We leave these questions to others, who will surely find many other problems to puzzle them and wonders to amaze and amuse them in this curious world of games. Only a certain feeling of incompleteness prompts us to add a final theorem.

THEOREM 100. *This is the last theorem in this book.*

(The proof is obvious.)

Epilogue

Reading this book for the first time in two decades has made me more aware of its defects than its merits—it's perhaps too obvious that it was written in a week! However, as a book it was an immediate success: so much so that the London Mathematical Society was able to use it to subsidize the other books of their Lecture Note Series in which it first appeared.

What has happened since then to the two new subjects it presented—the theory of Surreal Numbers and the additive theory of partizan games? Since the new edition of *Winning Ways* will describe the progess in additive game theory, I shall here concentrate on the Surreal Numbers, for which the answer is that there definitely has been some progress, but not enough. Please make some more!

The Surreal Numbers have been the topic of many research papers and a number of books. After Donald Knuth's *Surreal Numbers* came Harry Gonshor's *The Theory of Surreal Numbers*, Norman Alling's *Foundations of Analysis over Surreal Number Fields*, and Philip Ehrlich's *Real Numbers, Generalizations of the Reals, and Theories of Continua.* There have also been several special sessions devoted to the Surreal Numbers at meetings of the American Mathematical Society.

Most of the authors who have written about them have chosen to define surreal numbers to be just their sign-sequences. This has the great advantage of making equality be just identity rather than an inductively defined relation, and also of giving a clear mental picture from the start. However, I think it has probably also impeded further progress. Let me explain why.

The greatest delight, and at the same time, the greatest mystery, of the Surreal numbers is the amazing way that a few simple "genetic" definitions magically create a richly structured Universe out of nothing. Technically, this involves in particular the facts that each surreal number is repeatedly redefined, and that the functions the definitions produce are independent of form. Surely real progress will only come when we understand the deep reasons why these particular definitions have this property? It can hardly be expected to come from an approach in which this problem is avoided from the start?

The sign-sequence definition has also the failing that it requires a prior construction of the ordinals, which are in ONAG produced as particular cases of the surreals. To my mind, this is another symptom of the same problem, because the definitions that work universally should automatically render such prior constructions unnecessary. There is also a peculiar emphasis on the number ½ that is totally unnecessary—in ONAG {1/3 | 2/3} is just as good a definition of ½ as {0|1} is—and that I think serves to obscure the underlying structure.

I believe the real way to make "surreal progress" is to search for more of these "genetic" definitions and seek to understand their properties. The rest of this Epilogue will describe the few small steps that have been taken in this direction, and the large amount that still remains to be done.

The first edition contained the remark that "I do not believe that there is any natural definition, for instance, of the function x^y for infinite y. Nor does there seem to be any particular point in making these definitions." However, soon after that appeared, Simon Norton produced a definition of Surreal integration that led to an acceptable logarithm, while Martin Kruskal gave an independent definition of the exponential that turned out to be its inverse. Using these, we can of course define the analytic power x^y to be $\exp(y.\log(x))$.

Kruskal also showed that there was indeed some "point in making these definitions". Namely, he hopes to extend virtually all of classical analysis to the surreal (and surcomplex) numbers, and then to use this to solve the old problem of giving precise meanings to the sums of asymptotic series. We shall take the time to explain Kruskal's program in some detail, since there are difficulties which have caused him to postpone publishing his partial results on it.

Hard-line mathematicians have not yet given any general definitions for the values of series like the one that appears in Stirling's formula:

$$\text{Log}(x!) \sim x.\log(x) - x + \tfrac{1}{2}.\log(2.\text{pi}.x) + 1/12x + \ldots \qquad (*).$$

Here the series on the right converges for no real value of x.

Instead, they use "~" here to mean just that the terms up to and including that in $1/x^n$ provide an approximation to $\log(x!)$ that is valid to order $o(1/x^n)$.

Kruskal hopes to change all this! He remarks that Stirling's series does converge for infinite surreal x, so that an independent definition of $x!$ would make it meaningful to say that (*) did or did not hold with equality for all such x. In addition, he hopes to prove a metatheorem to the effect that when series like this converge to known functions for all infinite surreal x, then we will run into no contradictions by using this to define their exact values for real x.

Most "classical" functions are defined by ordinary differential equations, so Kruskal proposes to define what it means to be the solution of such an equation in the Surreal Realm, and to prove the necessary existence and uniqueness theorems. The archetypical case is the notion

$$\int_a^b f(t)dt$$

for "reasonable" functions f, but unfortunately a few years ago Kruskal found a problem with Norton's old definition of this.

Since that definition has never been published, and since it neatly illustrates several of the problems of finding genetic definitions, I shall briefly describe it here.

We suppose f is the function given by a 'genetic definition' $f = \{f^L \mid f^R\}$, where $f^L = f^L(x; x^L, x^R)$ and $f^R = f^R(x; x^L, x^R)$ will be functions of x and some Left and Right options x^L and x^R (of which there may be many).

Then the definition of

$$\int_a^b f(t)dt$$

is

$$\left\{ \begin{array}{c|c} \int_a^{b^L} f + D\int_{b^L}^b f^L, \int_a^{b^R} f - D\int_b^{b^R} f^R & \int_a^{b^L} f + D\int_{b^L}^b f^R, \int_a^{b^R} f - D\int_b^{b^R} f^L \\ \int_{a^R}^b f + D\int_a^{a^R} f^L, \int_{a^L}^b f - D\int_{a^L}^a f^R & \int_{a^R}^b f + D\int_a^{a^R} f^R, \int_{a^L}^a f - D\int_{a^L}^a f^L \end{array} \right\}$$

Each option in this is obtained by adding an integral of f over a simpler range (such as $[a, b^L]$) to a "dissected integral" of f^L or f^R. The typical such dissected integral, say

$$D\int_c^{c'} f^L,$$

is defined to be

$$\int_{c_0}^{c_1} f^{L_1}(t; c_0, c_1)dt + \int_{c_1}^{c_2} f^{L_2}(t; c_1, c_2)dt + \ldots + \int_{c_{n-1}}^{c_n} f^{L_n}(t, c_{n-1}, c_n)dt,$$

which depends on choices of a dissection $D = \{c = c_0, c_1, c_2, \ldots, c_n = c'\}$ of $[c, c']$ and of particular Left options $f^{L_1}, f^{L_2}, \ldots f^{L_n}$ for each subinterval of D. When integrating f^{L_i} over $[c_{i-1}, c_i]$, all the Left options of t are replaced by c_{i-1} and all the Right ones by c_i.

Norton and I showed that this definition correctly integrates polynomials, and also integrates $1/x$ to a satisfactory logarithm that is the functional inverse of Kruskal's exponential, but there are many problems with the definition in general.

The first of these is that the collections of options that appear in the definition are Proper Classes rather than sets. This, however, is not a serious concern, since in practice these Classes are dominated by sets.

A bigger worry is that the definition is "intensional", meaning that it depends on the way f is defined, and not just on its values, as do the standard "extensional" definitions of classical mathematics. There are examples of functions defined by two different definitions that have the same values everywhere, but integrate differently! However, these examples are rather artificial, and we hoped eventually to discover inatural restrictions which would make them go away.

Perhaps the most important problem was that we were never able to erect the mathematical theory necessary to show that (under suitable conditions) Norton's integral had the desired properties such as linearity, translation invariance, and so on. For twenty years we believed that nevertheless the definition was probably "correct" in some natural sense, and that these difficulties arose merely because we did not understand exactly which genetic definitions were "legal" to use in it.

Kruskal has now made some progress of a rather sad kind by showing that this belief was false. Namely, the definition integrates e^t over the range $[0, \omega]$ to the wrong answer e^ω, rather than $e^\omega - 1$, independently of whatever reasonable genetic definition we give for the exponential function.

In the quarter century that has elapsed since the first edition of this book was prepared, we have learned that (contrary to my opinion of that time) there are indeed natural definitions of x^y and some similar functions, but have also learned that our impressions about how to enlarge their number (by integration) were wrong. I still believe that a correct and natural theory will one day be found, but am unwilling to hazard a guess as to when this will be.

It is pleasant to be able to end this Epilogue on a more positive note. Jacob Lurie, who as a high school student won the Westinghouse competition with an essay about the Surreal numbers, has very recently proved my conjecture that the Group of all games is the universally embedding partially ordered Abelian Group. I am pleased to be able to congratulate him for the second time!

John Conway
12 October 2000

APPENDIX

Game	Pronunciation	Definition
0	nought *or* zero	$\{\,\|\,\}$
1	one	$\{0\,\|\,\}$
2	two	$\{1\,\|\,\} = \{0, 1\,\|\,\} = 1 + 1$
3	three	$\{2\,\|\,\} = \{0, 1, 2\,\|\,\} = 1 + 1 + 1$
$\frac{1}{2}$	half	$\{0\,\|\,1\}$
$\frac{1}{4}$	quarter	$\{0\,\|\,\frac{1}{2}\}$
$\frac{3}{4}$	three-quarters	$\{\frac{1}{2}\,\|\,1\}$
$1\frac{1}{2}$	one-and-a-half	$\{1\,\|\,2\}$
-1	minus-one	$\{\,\|\,0\}$
-2	minus-two	$\{\,\|\,-1\}$
$-\frac{1}{2}$	minus-half	$\{-1\,\|\,0\}$
...	etc.	
ω	omega	$\{0, 1, 2, 3, \ldots\,\|\,\}$
$\omega + 1$	omega-plus-one	$\{\omega\,\|\,\} = \{0, 1, \ldots, \omega\,\|\,\}$
$\omega.2$	omega-times-two	$\{0, 1, \ldots, \omega, \omega + 1, \ldots\,\|\,\} = \omega + \omega$
ω^2	omega-squared	$\{0, 1, \ldots, \omega, \ldots, \omega 2, \ldots, \omega 3, \ldots\,\|\,\}$
α	(the general ordinal)	$\{\ldots, \beta, \ldots (\beta < \alpha)\,\|\,\}$
$\omega - 1$	omega-less-one	$\{0, 1, 2, \ldots\,\|\,\omega\}$
$\frac{\omega}{2}$	half-omega	$\{0, 1, \ldots\,\|\,\omega, \omega - 1, \ldots\}$
$\sqrt{\omega}$	root-omega	$\{0, 1, \ldots\,\|\,\omega, \frac{\omega}{2}, \ldots\}$
$\frac{1}{\omega}$	omeg'th	$\{0\,\|\,1, \frac{1}{2}, \ldots\}$
$\frac{2}{\omega}$	two-omeg'ths	$\left\{\frac{1}{\omega}\,\middle\|\,1, \frac{1}{2}, \ldots\right\}$
$\frac{1}{2\omega}$	half-omeg'th	$\left\{0\,\middle\|\,\frac{1}{\omega}\right\}$

Game	Pronunciation	Definition
ω^x	omega-to-the-x	$\{0, r\omega^{x^L} \mid r\omega^{x^R}\}$ (r positive real).
x	(the general number)	{simpler numbers $< x$ | simpler numbers $> x$}
$-x$	minus x	$\{-x^R \mid -x^L\}$
$x + y$	x plus y	$\{x^L + y, x + y^L \mid x^R + y, x + y^R\}$
xy	x times y	$\{x^L y + xy^L - x^L y^L, x^R y + xy^R - x^R y^R \mid x^L y + xy^R - x^L y^R, x^R y + xy^L - x^R y^L\}$
$x{:}y$	x ordinal-sum y	$\{x^L, x{:}y^L \mid x^R, x{:}y^R\}$

The definitions of these operations apply also to other games. For the operation of inversion ($1/x$) see Chapter 1.

0	zero game	$\{\mid\} = *0$
$*$	star	$\{0 \mid 0\} = *1$
$*2$	star-two	$\{0, * \mid 0, *\}$
$*n$	star-n	$\{*0, *1, \ldots, *(n-1) \mid *0, *1, \ldots, *(n-1)\}$
$*\alpha$	star-α	$\{*\beta(\beta < \alpha) \mid *\beta(\beta < \alpha)\}$
$*abc\ldots$	(typical superstar)	$\uparrow * + \{*a, *b, *c, \ldots \mid *0, *1, *2, \ldots\}$
$\uparrow$	up	$\{0 \mid *\}$
$\uparrow *$	up-star	$\{0, * \mid 0\} = \uparrow + *$
$\uparrow *n$	up-star-n	$\{0 \mid *(n +_2 1)\} = \uparrow + *n$
$\Uparrow$	double-up	$\{0 \mid \uparrow *\} = \uparrow + \uparrow$
$\Uparrow *$	double-up-star	$\{0 \mid \uparrow\} = \uparrow + \uparrow + *$
$\downarrow$	down	$\{* \mid 0\} = -\uparrow$
$\Downarrow$	double-down	$\{\downarrow * \mid 0\} = -\Uparrow$
$\uparrow 2$	up-two	$\{\uparrow \mid *\} = \uparrow + \uparrow^2$
$\uparrow 3$	up-three	$\{\uparrow 2 \mid *\} = \uparrow + \uparrow^2 + \uparrow^3$
$\uparrow 2*$	up-two-star	$\{0, \uparrow * \mid 0\} = \uparrow 2 + *$
$\uparrow \frac{1}{2}$	up-half	$\{0 \mid \uparrow, *\}$
$\uparrow \frac{1}{2} *$	up-half-star	$\{0, * \mid 0, \uparrow *\} = \uparrow \frac{1}{2} + *$
$\uparrow x$	up-x	$\{*, \uparrow x^L \mid *, \uparrow x^R\} = \{* \mid x \mid *\}$ for numbers x
$\uparrow x*$	up-x-star	$\{0, \uparrow x^L * \mid 0, \uparrow x^R *\} = \{0 \mid x \mid 0\} = \uparrow x + *$
$\downarrow x$	down-x	$-(\uparrow x)$ (etc.).
$\Uparrow x$	double-up-x	$\{\mid x \mid \uparrow *\} = \uparrow + \uparrow x$ for $x \geqslant 1$.

Game	Pronunciation	Definition
$\Uparrow\frac{1}{2}$	double-up-half	$\{0 \mid \uparrow *, \Uparrow\} > \uparrow + \uparrow\frac{1}{2}$ (etc.)
$\Uparrow x; y$	double-up $x; y$	$\{0 \mid \uparrow x *, \uparrow y\}$
$\uparrow^2$	up-second	$\{0 \mid \downarrow *\}$
$\uparrow^3$	up-third	$\{0 \mid \downarrow 2 *\}$
$\downarrow_2$	down-second	$\{\uparrow * \mid 0\} = -(\uparrow^2)$ (etc.)
$\uparrow^x$	up-xth	by the formula $\uparrow x = \uparrow$ **On** $(1 - \uparrow^x)$
$x.\uparrow = \hat{x}$	x-fold up	$\{x^L + \Uparrow * \mid x^R + \Downarrow *\}$ (x not integral)
$*.\uparrow = \hat{*}$	starfold up	$\{\Uparrow * \mid \Downarrow *\}$
$\uparrow n^+$	up-n-plus	$\{\uparrow a^+ + \uparrow b^+, 0 \mid *\}$ $(a + b = n)$
$\uparrow^{n+}$	up-nth-plus	$\uparrow^{1+} + \uparrow^{2+} + \ldots + \uparrow^{n+} = \uparrow n^+$
$\uparrow^{2+}$	up-second-plus	$\{\uparrow \mid \downarrow *\}$
$\uparrow * n$	up-star-n	$\{0 \mid *(n +_2 1)\}$
$\hat{\frac{1}{2}}$	semi-up	$\{\Uparrow * \mid \downarrow *\}$
$\hat{\frac{3}{2}}$	sesqui-up	$\{(3.\uparrow) * \mid *\}$
$\hat{\frac{1}{2}} *$	semi-up-star	$\{\Uparrow \mid \downarrow\}$
$\frac{*}{2}$	semi-star	$\{*, \uparrow \mid \downarrow *, 0\}$
G	(general game)	$\{A, B, C, \ldots \mid D, E, F, \ldots\}$, if Left has moves to $A, B, C, \ldots$ and Right to $D, E, F, \ldots$.
G^L	Left option	one of $A, B, C, \ldots$ in above (so $G = \{G^L \mid G^R\}$)
G^R	Right option	one of $D, E, F, \ldots$ in above (so $G = \{G^L \mid G^R\}$)
$A \mid B$	A slash B	abbreviates $\{A \mid B\}$
$\{A, B, \ldots\}$	set $A, B, \ldots$	abbreviates $\{A, B, \ldots \mid A, B, \ldots\}$
$\pm X$	plus or minus X	abbreviates $\{X \mid -X\}$
$\pm(X, Y, \ldots)$	(similarly)	abbreviates $\{X, Y, \ldots \mid -X, -Y, \ldots\}$
$\{G^L \mid X \mid G^R\}$ or $G{:}X$	G-sum-x	the function $f(X)$ defined by $f(X) = \{G^L, f(X^L) \mid G^R, f(X^R)\}$
$+_x$	tiny-x	$\{0 \mid \{0 \mid -x\}\}$
$A \Vert B \mid C$	A slashes B slash C	abbreviates $\{A \mid \{B \mid C\}\}$ (etc.)
$+_2$	tiny-two	$\{0 \mid \{0 \mid -2\}\}$
∞	∞	$\{\mathbb{R} \mid \{\mathbb{R} \mid \mathbb{R}\}\}$
$\pm\infty$	plus or minus ∞	$\{\mathbb{R} \mid \mathbb{R}\} = \{\infty \mid -\infty\}$
$2\infty = \infty_0$	twice infinity	$\{\mathbb{R} \Vert \mathbb{R} \mid 0\}$

Game	Pronunciation	Definition
∞^{-1}	infiniteth	$\{1 \mid \mathbb{R}^+ \parallel \mathbb{R}^+ \mid\mid\mid \mathbb{R}^+\}$
$\uparrow(\pm 1)$	up plus-or-minus one	$\{\uparrow, * \mid \downarrow, *\}$
$\uparrow 2^+ *$	up two-plus star	$\{\Uparrow * \mid 0\}$
$\uparrow^{2+} *$	up second-plus star	$\{0, \uparrow * \mid \downarrow\}$
$\uparrow 2 \uparrow^2$	up two-seconds	$\{0 \mid \uparrow^2 *\}$
$\uparrow 2 \uparrow^2 *$	up two-seconds star	$\{0 \mid \uparrow^2\}$
$\uparrow^2 *$	up-second star	$\{0, * \mid \downarrow\}$
$\uparrow\downarrow_2$	up down-second	$\{\Uparrow * \mid *, \uparrow\}$
$\uparrow\downarrow_2 *$	up down-second star	$\{\Uparrow \mid 0, \uparrow *\}$

REFERENCES

Not all of these are mentioned in the text, the others being on related subjects and particular games. The reader should be warned that the name "Theory of Games" usually refers to a theory of a different kind used in economics and political decision-making. The series *Contributions to the Theory of Games*, edited by Tucker and others, in the Princeton Annals of Mathematics Studies, nos 24, 28, 39, 40 contains many papers on this subject, and the fourth volume has a large bibliography.

1. C. L. Bouton, *Nim, a game with a complete mathematical theory*, Ann. Math., Princeton (2), 3(1902), 35–39.
2. J. H. Conway, *All games bright and beautiful*, The University of Calgary Math. Research Paper #295. October 1975.
3. J. H. Conway, *All numbers, great and small*, The Univ. of Calgary Math. Research Paper #149, February, 1972.
4. Aviezri S. Fraenkel, *Combinatorial games with an annihilation rule*, Proc. A.M.S. Symp. App. Math., 20(1974), 87–91.
5. David Gale, *A curious nim-type game*, Amer. Math. Monthly, 81(1974), 876–879.
6. Solomon W. Golomb, *A mathematical investigation of games of "take-away"*, J. Combinatorial Theory, 1(1966), 443–458.
7. P. M. Grundy, *Mathematics and games*, Eureka, 2(1939), 6–8.
8. P. M. Grundy and C. A. B. Smith, *Disjunctive games with the last player losing*, Proc. Camb. Philos, Soc., 52(1956), 527–533.
9. R. K. Guy and C. A. B. Smith. *The G-values for various games*, Proc. Cambridge Philos. Soc., 52(1956), 514–526.
10. Olof Hanner, *Mean play of sums of positional games*, Pacific J. Math., 9(1959), 81–99; M.R. 21 #3277.
11. P. G. Hinman, *Finite termination games with tie*, Israel J. Math., 12(1972), 17–22.
12. John C. Holladay, *Cartesian products of termination games*. Contributions to the theory of games, vol. 3, 189–200. #39 Ann. Math. Stud. Princeton. M.R. 20 #2236, 1957.
13. John C. Holladay, *Matrix Nim*, Amer. Math. Monthly, 65(1958), 107–109; M.R. 20 #4812.
14. Ja. G. Kljušin, *Equivalence theorems for general positional games* (Russian) in Advances in Game Theory (Proc. 2nd All-Union Conf. on Game Theory, Vilnius, 1971), 209–211 Zdat "Mintis" Vilnius 1973.
15. Donald E. Knuth, *Surreal numbers, Addison-Wesley*, 1974.

16. John Milnor, *Sums of positional games*, Annals of Math. Study # 28 (Kuhn & Tucker), Princeton 1953, 291–301.
17. David Singmaster, *Almost all games are first person games*,
18. C. A. B. Smith, *Graphs and composite games*, J. Combinatorial Theory, 1(1966), 51–81.
19. R. P. Sprague, *Über mathematische Kampfspiele*, Tôhoku Math. J., 41(1935–6), 438–444; Zbl. 13, 290.
20. ———, *Recreation in mathematics* (trans. T. H. O'Beirne) Blackie, 1963, # 24 Odd is the winner, pp. 18,
21. H. Steinhaus, *Definitions for a theory of games and pursuit*, Mýsl. Akad. Lwów 1, # 1(1925), 13–14; reprinted in Naval Res. Logist. Quart. 7(1960), 105–108.

The principal books and papers published about Surreal Numbers since the first edition of *On Numbers and Games* are:

Berarducci, Alessandro: *Factorization in generalized power series.* Trans. Am. Math. Soc. 352, No.2, 553-577 (2000).

Beyer, W.A. and Louck, J.D.: *Transfinite function iteration and surreal numbers.* Adv. Appl. Math. 18, No.3, 333-350, Art. No.AM960513 (1997).

Conway, John Horton and Guy, Richard K.: *The book of numbers.* Berlin: Springer-Verlag. 1996.

Ehrlich, Philip (ed.): *Real numbers, generalizations of the reals, and theories of continua.* Synthese Library. 242. Dordrecht: Kluwer Academic Publishers. 1994.

Gardner, Martin: *Penrose tiles to trapdoor ciphers ... and the return of Dr. Matrix. Rev. ed.* Washington, DC: The Mathematical Association of America. 1997.

Gonshor, Harry: *An Introduction to the theory of surreal numbers*, Lond. Math. Soc. Lect. Note Ser. 110, Cambridge University Press, 1986 (Zbl 595.12017).

Lemire, Denis: *Decompositions successives de la forme normale d'un surreel et generalisation des ε-nombres.* Ann. Sci. Math. Que. 21, No.2, 133-146 (1997).

Louck, James D.: *Conway numbers and iteration theory.* Adv. Appl. Math. 18, No.2, 181-215 (1997).

Lurie, Jacob: *The effective content of surreal algebra.* J. Symb. Log. 63, No.2, 337-371 (1998).

Index

He writes indexes to perfection.
Oliver Goldsmith, Citizen of the World (letter 29)